T0313342

Advanced Polymeric Systems:
Applications in Nanostructured Materials, Composites and Biomedical Fields

RIVER PUBLISHERS SERIES IN POLYMER SCIENCE

Series Editors:

SAJID ALAVI
Kansas State University
USA

YVES GROHENS
University of South Brittany
France

SABU THOMAS
Mahatma Gandhi University
India

Indexing: all books published in this series are submitted to the Web of Science Book Citation Index (BkCI), to SCOPUS, to CrossRef and to Google Scholar for evaluation and indexing

The "River Publishers Series in Polymer Science" is a series of comprehensive academic and professional books which focus on theory and applications of Polymer Science. Polymer Science, or Macromolecular Science, is a subfield of materials science concerned with polymers, primarily synthetic polymers such as plastics and elastomers. The field of polymer science includes researchers in multiple disciplines including chemistry, physics, and engineering.

Books published in the series include research monographs, edited volumes, handbooks and textbooks. The books provide professionals, researchers, educators, and advanced students in the field with an invaluable insight into the latest research and developments.

Topics covered in the series include, but are by no means restricted to the following:

- Macromolecular Science
- Polymer Chemistry
- Polymer Physics
- Polymer Characterization

For a list of other books in this series, visit www.riverpublishers.com

Advanced Polymeric Systems:
Applications in Nanostructured Materials, Composites and Biomedical Fields

Editors

Didier Rouxel
Institut Jean Lamour, Université de Lorraine, France

K. M. Praveen
Muthoot Institute of Technology & Science (MITS), India

Indu Raj
Government Dental College,
International and Mahatma Gandhi University, India

Sandhya Gopalakrishnan
Government Dental College, Mahatma Gandhi University, India

Nandakumar Kalarikkal
School of Pure and Applied Physics, Mahatma Gandhi University, India

Sabu Thomas
Mahatma Gandhi University, India

River Publishers

Routledge
Taylor & Francis Group
LONDON AND NEW YORK

Published 2020 by River Publishers
River Publishers
Alsbjergvej 10, 9260 Gistrup, Denmark
www.riverpublishers.com

Distributed exclusively by Routledge
4 Park Square, Milton Park, Abingdon, Oxon OX14 4RN
605 Third Avenue, New York, NY 10017, USA

Advanced Polymeric Systems: Applications in Nanostructured Materials, Composites and Biomedical Fields / by Didier Rouxel, K. M. Praveen, Indu Raj, Sandhya Gopalakrishnan, Nandakumar Kalarikkal, Sabu Thomas.

© 2020 River Publishers. All rights reserved. No part of this publication may be reproduced, stored in a retrieval systems, or transmitted in any form or by any means, mechanical, photocopying, recording or otherwise, without prior written permission of the publishers.

Routledge is an imprint of the Taylor & Francis Group, an informa business

ISBN 978-87-7022-136-8 (print)

While every effort is made to provide dependable information, the publisher, authors, and editors cannot be held responsible for any errors or omissions.

Contents

2   Fabrication of Natural Dye-Sensitised Solar Cells Based on Quasi Solid State Electrolyte Using TiO$_2$ Nanocomposites   **31**
N. Suganya, G. Hari Hara Priya and V. Jaisankar

5   A Comparative Approach to Structural Heterogeneity of Polyaniline and Its ZnO Nanocomposites   85

Bhabhina Ninnora Meethal, P. C. Ajisha,
Dharsana M. Vidyadharan, Jyothilakshmi V. Prakasan
and Sindhu Swaminathan

III   Bio-polymers   101

6   Synthesis and Characterisation of Polyurethanes from Bio-Based Vegetable Oil   103

D. Venkatesh and V. Jaisankar

7   Application of *Lepidium sativum* as an Excipient in Pharmaceuticals   113

S. V. Sutar, S. S. Shelake, S. V. Patil and S. S. Patil

8   Role of Polyhydroxyalkanoates (PHA-biodegradable Polymer) in Food Packaging   135

Abhishek Dutt Tripathi, Simmie Sebstraien,
Kamlesh Kumar Maurya, Suresh Kumar Srivastava,
Shankar Khade and Kundan

Preface

Recent advances in polymer science have led to the generation of high quality materials for various applications in day-to-day life. The incorporation of various macro-, micro- and nano-sized fillers into polymers has shown strong potential in generating novel polymer material with improved properties. This book covers the latest advancements in the field of polymer nanocomposites and polymer composites for varied applications.

The major topics discussed in the book include:

- Nanostructured materials for energy applications
- Nanostructured polymer composites
- Bio-polymers
- Nanostructured polymers for biomedical applications

Chapter 1 discusses different phase change materials (PCMs) for latent heat storage systems. The various supporting materials and additives used for preparing nanocomposite PCMs have been elaborated. Also, the routes of preparation have been detailed in this chapter. The varieties of materials presented have highly porous nature, high thermal energy storage density and high thermal conductivity.

Chapter 2 explores the possibility of natural dye sensitized solar cells (NDSSCs) as efficient and low cost photovoltaic cells. In this chapter, the preparation of TiO_2, TiO_2- ZnO, TiO_2-CuO and polymer gel electrolyte were discussed. The effects of these oxide layers on the performance of DSSCs assembled with a gel polymer electrolyte were investigated. The prepared core/shell nanostructures and gel polymer electrolytes were characterized by various analytical methods and results discussed.

Chapter 3 has focused on developing ZnO nanomaterial and applying it for solar cell application. ZnO nanomaterials were developed in two different ways and were incorporated with organic materials to form hybrid type device. Conditions for the deposition of ZnO nanorods, device fabrication using ZnO nanorods and its modification using PEDOT:PSS layer and Eu doping were explored.

Chapter 4 features the effects of nanosilica of size upto 60–80 nm as inorganic nanofillers for enhancement of mechanical properties for PP/SEBS blends. Different compositions of PP/SEBS/Si blend nanocomposites were formulated and prepared by melt mixing method. Effect of nanosilica on the mechanical, viscoelastic and morphological properties of PP/SEBS blend has been investigated and the results discussed.

Chapter 5 – In this chapter, polyaniline-zinc oxide hybrid composites were prepared through an easy one pot facile method following two different protocols. Various characterisation techniques were used to investigate the structural and morphological differences between these two nanocomposites. The reactive oxygen species (ROS) generated from the nanocomposites were quantified and compared. The photocatalytic mechanism responsible for the dye degradation process has been explained in detail in this chapter.

Chapter 6 discusses the importance of Bio-Renewable Polyurethanes (PUs)as one of the most versatile class of high performance industrial material. In the present investigation, the polyurethanes were prepared by polyols (SCOL) and isocyanates by polycondensation reaction. Vegetable oil based polyurethanes have been prepared by reacting polyols and polyurethanes and various results discussed.

Chapter 7 explores the effectiveness of Lepidium Sativumas a binder for pharmaceutical formulations. Aspirin & Ibuprofen tablets were prepared by wet granulation technique using Lepidium Sativum as a tablet binder. The prepared tablets were evaluated for physiochemical characteristics and the binding efficacies of the Lepidium Sativum were discussed.

Chapter 8 is a review featuring the biodegradable polymers as a suitable alternative to plastics. The production of Polyhydroxyalkanoates (PHA-biodegradable polymer) can be done using substrates that are waste from any food industry or even municipal wastewater solids can be used for the accumulation of PHAs. Role of PHA in food packaging is explored in detail.

Chapter 9 discusses a novel method of production of Xylitol, which is a natural polyol and most widely known for its sugar substitute properties in diabetic patients. In this present investigation, xylitol production by *Candida parapsilosis* strain is inverstigated and reported.

Chapter 10 is a review on the role of natural biopolymers such as polysaccharides in controlled therapeutics especially for cancer chemotherapy. Nanocellulose, based polysaccharide self-assembly is expected to lead into various morphologies opening up a novel strategy for controlled and targeted drug delivery with minimum side effects.

Chapter 11 explains the green syntheses of silver nanoparticles which is an emerging area. This technique is nontoxic, cost effective and environment friendly. Green plants contain several biomolecules such as alkaloids, flavonoids, terpenes etc. which acts as reducing as well as capping agents. The present study reported an extracellular rapid biosynthesis of silver nanoparticles from $AgNO_3$ using aqueous leaf extract of *Pimentadioica*and various results discussed.

List of Contributors

A. P. Meera, *Research and Post Graduate Department of Chemistry & Polymer Chemistry, KSMDB College, Sasthamcotta, Kollam, Kerala, India; E-mail: apmeera@gmail.com; meeradbc@gmail.com*

Aarsha Surendren, *Materials Science and Technology Division, Council of Scientific and Industrial Research – National Institute for Interdisciplinary Science and Technology, Thiruvananthapuram, India*

Abhishek Dutt Tripathi, *Centre of Food Science and Technology, Institute of Agricultural Sciences, Banaras Hindu University, Uttar Pradesh, India; E-mail: abhi_itbhu80@rediffmail.com*

Asha Susan Chacko, *Materials Science and Technology Division, Council of Scientific and Industrial Research – National Institute for Interdisciplinary Science and Technology, Thiruvananthapuram, India*

Asit B. Samui, *Department of Polymer and Surface Engineering, Institute of Chemical Technology, Mumbai, India; E-mail: absamui@gmail.com*

Avinash Nath Tiwari, *Department of Mechanical Engineering, Jagannath University, Jaipur, Rajasthan, India*

Balakrishnaraja Rengaraju, *Department of Biotechnology, Bannari Amman Institute of Technology, Sathyamangalam, India; E-mail: balakrish-narajar@bitsathy.ac.in*

Bhabhina Ninnora Meethal, *Department of Nanoscience and Technology, University of Calicut, Kerala, India*

D. Venkatesh, *PG & Research Department of Chemistry, Presidency College (Autonomous), Chennai, Tamil Nadu, India*

D. Vinotha, *Department of Biotechnology, Bannari Amman Institute of Technology, Sathyamangalam, India*

Dharsana M. Vidyadharan, *Department of Nanoscience and Technology, University of Calicut, Kerala, India*

G. Hari Hara Priya, *Department of Chemistry, SDNB Vaishnav College for Women, Chromepet, Chennai, India*

Harekrishna Panigrahi, *Department of Mechanical Engineering, Jagannath University, Jaipur, Rajasthan, India; E-mail: harekrishnapanigrahi91@gmail.com*

Jyothilakshmi V. Prakasan, *Department of Nanoscience and Technology, University of Calicut, Kerala, India*

K. P. Vijayakumar, *Thin Film Photovoltaic Division, Department of Physics, Cochin University of Science and Technology, Cochin, India; E-mail: vijayakumar.kandathil@gmail.com*

K. S. Ranjith, *Advanced Materials and Devices Laboratory, Department of Physics, Bharathiar University, Coimbatore, India; E-mail: ranjuphy@gmail.com*

Kamlesh Kumar Maurya, *Centre of Food Science and Technology, Institute of Agricultural Sciences, Banaras Hindu University, Uttar Pradesh, India; E-mail: kamleshcfstbhu@gmail.com*

Kundan, *School of Biochemical Engineering, Indian Institute of Technology (BHU), Varanasi, Uttar Pradesh, India; E-mail: kunal786singh@gmail.com*

M. P. Singh, *Department of Mechanical Engineering, JECRC University, Jaipur, Rajasthan, India*

N. Suganya, *PG and Research Department of Chemistry, Presidency College (Autonomous), Chennai, India*

P. Anju, *Materials Science and Technology Division, Council of Scientific and Industrial Research – National Institute for Interdisciplinary Science and Technology, Thiruvananthapuram, India*

P. B. Sreelekshmi, *Research and Post Graduate Department of Chemistry & Polymer Chemistry, KSMDB College, Sasthamcotta, Kollam, Kerala, India*

P. C. Ajisha, *Department of Nanoscience and Technology, University of Calicut, Kerala, India*

P. Ramalingam, *Department of Biotechnology, Kumaraguru College of Technology, Coimbatore, India*

R. Geethu, *Thin Film Photovoltaic Division, Department of Physics, Cochin University of Science and Technology, Cochin, India; E-mail: geethumangalath@gmail.com*

R. T. Rajendra Kumar, *Department of Nanoscience and Technology, Bharathiar University, Coimbatore, India; E-mail: rtrkumar@buc.edu.in*

Reshma R. Pillai, *Research and Post Graduate Department of Chemistry & Polymer Chemistry, KSMDB College, Sasthamcotta, Kollam, Kerala, India*

S. S. Patil, *Department of Pharmaceutics, Ashokrao Mane College of Pharmacy, Peth-Vadgaon, Hatkanangale, Kolhapur, Maharashtra, India*

S. S. Shelake, *Department of Pharmaceutics, Ashokrao Mane College of Pharmacy, Peth-Vadgaon, Hatkanangale, Kolhapur, Maharashtra, India*

S. V. Patil, *Department of Pharmaceutics, Shree Santkrupa College of Pharmacy, Ghogaon, Karad, Satara, Maharashtra, India*

S. V. Sutar, *Department of Pharmaceutical chemistry, Ashokrao Mane College of Pharmacy, Peth-Vadgaon, Hatkanangale, Kolhapur, Maharashtra, India*

Shankar Khade, *School of Biochemical Engineering, Indian Institute of Technology (BHU), Varanasi, Uttar Pradesh, India; E-mail: khadeshankar007@gmail.com*

Simmie Sebstraien, *Centre of Food Science and Technology, Institute of Agricultural Sciences, Banaras Hindu University, Uttar Pradesh, India; E-mail: shimmy.sebastian94@gmail.com*

Sindhu Swaminathan, *Department of Nanoscience and Technology, University of Calicut, Kerala, India; E-mail: sindhus@uoc.ac.in*

Smrutirekha Mishra, *School of Chemical Technology, Kalinga Institute of Industrial Technology (KIIT), Bhubaneswar, Odisha, India*

Sumesh Soman, *Materials Science and Technology Division, Council of Scientific and Industrial Research – National Institute for Interdisciplinary Science and Technology, Thiruvananthapuram, India*

Suresh Kumar Srivastava, *School of Biochemical Engineering, Indian Institute of Technology (BHU), Varanasi, Uttar Pradesh, India; E-mail: sksrivastava.bce@itbhu.ac.in*

Swati Sundararajan, *Zuckerberg Institute of Water Research, Jacob Blaustein Institutes for Desert Research, Ben-Gurion University of the Negev, Sede-Boqer Campus, Israel; E-mail: swati14ssr@gmail.com*

V. Jaisankar, *PG & Research Department of Chemistry, Presidency College (Autonomous), Chennai, Tamil Nadu, India; E-mail: vjaisankar@gmail.com*

V. S. Prasad, *Materials Science and Technology Division, Council of Scientific and Industrial Research – National Institute for Interdisciplinary Science and Technology, Thiruvananthapuram, India; E mail: vsprasad@niist.res.in*

List of Figures

List of Tables

List of Abbreviations

η	Efficiency
AC	Activated carbon
AFM	Atomic Force Microscopy
Ag NP	Siliver nanoparticle
ASTM	American Society for Testing & Materials
BP	British Pharma
Brij 58	Polyethylene glycol hexadecyl ether
CBD	Chemical bath deposition
CMK-5	Ordered mesoporous carbon
CNIC	Carbon nitride intercalation compound
CNS	Carbon nanosphere
CNT	Carbon nanotube
CNTS	Carbon nanotube sponge
CSP	Chemical spray pyrolysis
EDTA	Ethylenediamine tetraacetic acid
EG	Expanded graphite
FF	Fill factor
FTIR	Fourier transform Infrared Spectrum
Gc	Garden cress
GCMS	Gas chromatography- mass spectrometry
GF	Graphite foam
GNP	Graphene nanoplatelets
GO	Graphene oxide
Has	Hydroxyalkanoates
HCl	Hydrochloric acid
HGA	Hybrid graphene aerogel
HGF	Hierarchical graphene foam
HMTA	Hexamine
HNT	Hallotsite nanotube
HPLC	High Performance Liquid Chromatography
HTPB	Hydroxyl terminated polybutadiene

I_0	Saturation current
I_L	Photo current
I_{sc}	Short circuit current
IP	Indian Pharmacopoeia
ITO	Indium Tin Oxide
I-V	Current–voltage
LAS	Alkylbenzene sulfonic acid
LCL	long chain length
MCL	medium-chain-length
MWCNT	Multiwalled carbon nanotube
NASA	National Aeronautics and Space Administration
NMR	Nuclear Magnetic Resonance
P_m	Maximum power point
P3HT	Poly(3-hex yl thiophene)
PAGF	Porous Al2O3@graphite foam
PCBM	[6,6]-phenyl-C_{60}-butyric acid methyl ester
PCMs	Phase change materials
PEDOT:PSS	Poly(3,4-ethylenedioxythiophene):polystyrene sulfonate
PEG	Polyethylene glycol
PHAs	poly hydroxyl alkanoates
PHB	Poly hydroxybutyrate
PHV	Polyhydroxyvalerate
PMMA	Polymethyl methacrylate
PU	Polyurethane
PV	Photovoltaics
PVP	Polyvinyl Pyrrolidine
PW	Paraffin wax
r^2	Regression coefficient
rGO	Reduced graphene oxide
rmp	Rotation per minute
RMS	Radial mesoporous silica
RMS	Root Mean Square
SA	Stearic acid
SCL	Short chain length
SD	Standard Deviation
SDS	Sodium dodecyl sulphate
SEM	Scanning Electron Microscopy
SG	Sulphonated graphene
SSPCMs	Solid state phase change materials

SWCNT	Singlewalled carbon nanotube
TEOS	Tetraethylorthosilicate
TES	Thermal energy storage
TR	Tangled root
USP	United States Pharmacopoeia
V_{oc}	Open circuit voltage
Voltage – J-V	Current density
XRD	X-ray diffraction
ZnO	Zinc Oxide
ZnO:Al	Aluminium doped ZnO

Part I

Nanostructured Materials for Energy Applications

1

Smart Nano-Enhanced Organic Phase Change Materials for Thermal Energy Storage Applications

Swati Sundararajan[1] and Asit B. Samui[2]

[1]Zuckerberg Institute of Water Research, Jacob Blaustein Institutes for Desert Research, Ben-Gurion University of the Negev, Sede-Boqer Campus, Israel
[2]Department of Polymer and Surface Engineering, Institute of Chemical Technology, Mumbai, India
E-mail: swati14ssr@gmail.com; absamui@gmail.com

Graphical Abstract

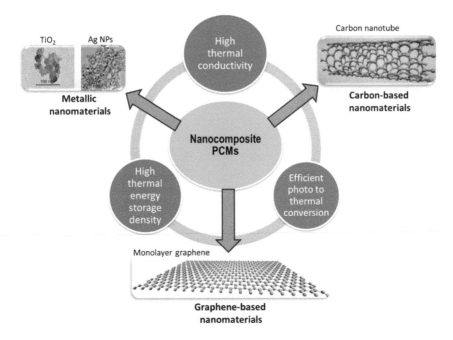

3

Among various sources of renewable energy, thermal energy storage has gained a wide popularity to bridge the gap between energy demand and energy generation. Latent heat storage systems, viz. phase change materials (PCMs), are most effective due to reversible high thermal energy storage density within a short temperature span. The organic PCMs suffer from disadvantages of low charging and discharging rates which in effect relates to the thermal conductivity. To enhance this thermal conductivity, nanocomposite PCMs have become hot targets of research to prepare high-performance PCMs. Various supporting materials and additives used for preparing nanocomposite PCMs along with routes of preparation and their performance have been detailed in this chapter. This chapter has been divided into different sections based on the nanomaterial used to enhance the performance of organic and inorganic PCMs followed by concluding remarks. The various sections are based on inorganic, organic (mostly based on nanocarbons) and organometallic nanomaterials.

Keywords: Phase Change Materials, Nanocomposites, Metallic nanoparticles, Carbon nanomaterials, Graphene nanoplatelets, Graphene foam

1.1 Introduction

The ever-rising energy demands with increasing population, depletion of fossil fuels and industrialisation has triggered the search for various renewable energy sources. These issues have created a need for energy security and efficient energy utilisation following which extensive research is going on in the field of renewable energy resources. The mismatch between energy supply and demand along with the utilisation of intermittent sources of energy can be filled in by one such system, thermal energy storage (TES) system [1]. TES can be pursued in the form of sensible heat, latent heat and thermochemical heat [2]. Sensible heat is stored with increasing or decreasing temperature of the system, utilising specific heat capacity of the material. Latent heat is stored when the material undergoes phase transformation under isothermal conditions. Latent heat storage materials, also known as phase change materials (PCM), absorb energy with phase transformation during heating and retrieve this energy during cooling process. During heating, the crystals melt by absorbing heat and during cooling, the crystals are reformed by releasing the absorbed heat. Thermochemical energy is stored by reversible chemical reactions due to the breaking and formation of molecular bonds. PCMs are most effective among TES

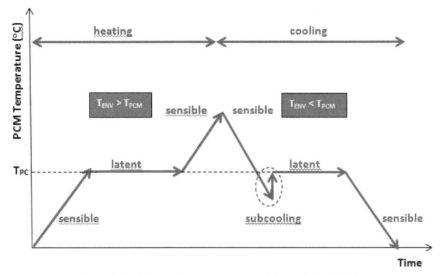

Figure 1.1   Schematic representation of the working of PCM.

materials due to high energy storage density within a narrow temperature range. The schematic of PCM working has been presented in Figure 1.1. The heat storage capacity of a PCM can be determined by using Equation 1.1 [3]:

$$Q = \int_{T_i}^{T_m} mC_p dT + ma_m \triangle H_m + \int_{T_m}^{T_f} mC_p dT \qquad (1.1)$$

A phase change material should have the following characteristics for maximum efficiency [3,4]:

Thermal:

- High thermal energy storage density
- Suitable phase change temperature
- High thermal conductivity
- Cycling stability

Physical:

- High density
- Little or no subcooling

- Low vapour pressure
- Small volume change

Chemical:

- Long-term stability
- Compatibility with container material
- Non-toxic, non-flammable

Economic:

- Abundance
- Availability
- Cost-effective
- Recyclability

PCMs were first employed by NASA for preparing thermoregulating suits of astronauts to protect from temperature fluctuations. PCMs find applications in many sectors which include thermoregulating textiles, smart buildings, active and passive solar energy storage, cooling of electronics, therapeutic packs, temperature-controlled greenhouses and waste heat recovery [5,6]. A number of novel applications have been explored recently which includes nanomedicines, biometric identification, anti-icing coating and thermal barcoding [7,8].

1.1.1 Types of PCM

PCMs can be classified based on the nature of the material as inorganic, organic and eutectic PCMs [9]. Inorganic PCMs consisting of hydrated salts and metallics have high thermal energy storage density and high thermal conductivity. But the disadvantages of inorganic PCMs, such as high degree of subcooling, incongruent melting, phase separation with repeated cycling and corrosion largely limit their applicability. On the other hand, organic PCMs are characterised by high latent enthalpy, little or no subcooling, self-nucleating behaviour, low vapour pressure, non-corrosive nature, good chemical and thermal stability. Organic PCMs consist of paraffin waxes (PW), fatty acids, esters, alcohols, sugars and poly(ethylene glycol). Eutectics can be of the inorganic or organic type using mixtures of two or more PCMs. Though these eutectics offer good phase change enthalpy, they melt incongruently. Also, extensive research is required to find eutectics with suitable mass ratio, phase stability and proper transition temperatures.

1.1.2 Physical Form of PCM

Phase change can be in the form of solid-liquid, solid-solid, liquid-gas and solid-gas [9,10]. Solid-gas and liquid-gas transformations have high latent enthalpy values but the volume changes associated with phase transition make the containment difficult and unfeasible for practical purposes. Solid-liquid transitions are economically attractive but containment is required in the liquid phase to prevent leakage due to large volume change. Solid-solid phase changes have lower enthalpies than solid-liquid phase change but offer many advantages of encapsulation and containment, small volume change, absence of liquid leakage and ability to process in desired shapes. The SSPCMs can be prepared by using both physical and chemical methods [11]. Form-stable or shape-stabilised composites can be developed by physical bonding of soft segment with a higher melting polymeric substrate such that the soft segment is embedded in the polymer matrix preventing its leakage during the phase transition from solid to liquid phase, as long as the temperature is maintained below the melting point of the polymer matrix. The physical methods include blending, adsorption and soaking. An inorganic/organic porous matrix is also used for making SSPCMs. Various chemical methods such as grafting, cross-linking and copolymerisation can also be utilised for the preparation of solid-solid PCMs, wherein the working material is chemically linked to the supporting material via different linkages such as urethane, epoxy, ester and ether. The chemical methods of modification yield more chemically and thermally stable PCMs but compromise the TES capacity.

However, the commercialisation of solid-solid PCMs is hindered due to low thermal conductivity of the composite PCMs, low storage density and leakage at temperatures beyond the melting temperature for form stable PCM [12]. If the thermal conductivity is low, then it would result in slow thermal storing and releasing properties and therefore, exhibit low thermal energy utilisation efficiency. The low thermal conductivity can be enhanced by two methods; using a thermally conductive supporting matrix or thermally conductive fillers [13]. With the progress of the development of conducting fillers, much higher performance is observed with a nano-supporting matrix or nanofillers. Thus, nanocomposite PCMs have become popular due to the advantages of direct applicability with high thermal energy storage density and high charging/discharging rates. The use of nanomaterials in PCMs has been illustrated schematically in Figure 1.2. Various nanomaterials such as metallic nanoparticles, carbon nanomaterials, graphene and derivatives have been used effectively to prepare high-performance phase change composites.

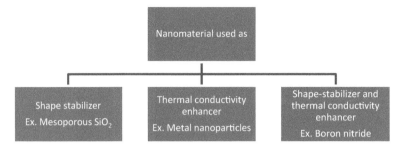

Figure 1.2   Schematic to illustrate the use of nanomaterials.

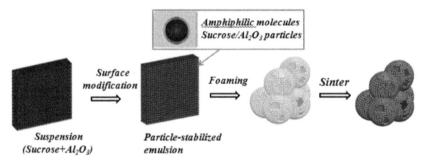

Figure 1.3   Preparation of porous alumina@graphite foam by particle-stabilised direct foaming method. Reprinted (adapted) with permission from Ref. [15]. Copyright (2017) American Chemical Society.

1.2 Inorganic Nanocomposites

Various inorganic nanocomposites can be prepared by utilising inorganic materials as supporting matrices for solid-liquid PCM.

Paraffin, a simple and versatile PCM, forms form-stable composites with hexagonal boron nitride sheets with a diameter of 0.5–2 μm and thickness of almost 100 nm [14]. The resulting heat transfer rate can be enhanced by 25% with about 10% loading of boron nitride. As a result of confinement, the latent enthalpy reduces by about 12% to 177 J.g^{-1}. Porous Al$_2$O$_3$@graphite foam (PAGF) has been prepared by using particle-stabilised direct foaming method as seen in Figure 1.3 [15]. Three-dimensional interpenetrating pore structure with 66 wt% PW is prepared to achieve fusion enthalpy of 105.8 J.g^{-1}. By using alumina foam, the conductivity of composite can be enhanced by 3.62 times to that of pure PW.

PEG can be loaded with 4 wt% GO as supporting material and 30 wt% boron nitride is added as thermally conductive filler. The incorporation of a high concentration of boron nitride can increase the thermal conductivity by 900% to a value of 3 W.m^{-1}.K^{-1}. At the same time, the photo-absorption is also improved [16]. The fusion enthalpy of 107.4 J.g^{-1} is not adversely affected by forming such kind of composite.

Flower-like TiO$_2$ nanoparticles with high surface area and unique pore structure can be used to encapsulate PEG up to 50.2 wt% [17]. The nanocomposite has good thermal reliability up to 200 thermal cycles with enthalpy of 86 J.g^{-1}. Radial mesoporous silica (RMS), prepared with cetyltrimethylammonium bromide template and tetraethylorthosilicate (TEOS) as SiO$_2$ precursor, can be vacuum impregnated with PEG [12]. Immersion time of 50 minutes with an immersion temperature of 70°C is optimum for the preparation of this PCM. Maximum enthalpy of 130 J.g^{-1} is possible by using 80 wt% PEG/RMS composite. Simultaneously, the supercooling is reduced by 4.8% - 19.7%.

Bulk carbon nitride (bulk-C$_3$N$_4$) and carbon nitride intercalation compounds (CNIC) are used as shape-stabilisers for PEG by using blending and impregnation method [18]. However, the PEG/C$_3$N$_4$ blend does not have any crystalline peaks of PEG due to the hindrance to its crystallisation. The PEG/CNIC blends having a porous structure, exhibit PEG diffraction peaks with lower intensity than that of PEG. The higher specific area of CNIC than that of bulk C$_3$N$_4$ aids the crystallisation of PEG. The PEG/bulk-C$_3$N$_4$ and PEG/CNIC composites could accommodate a maximum of 40 wt% and 60 wt% PEG. Reduction of melting temperature by 24°C from that of pure PEG is exhibited by the PEG/CNIC blend having a fusion enthalpy of 46 J.g^{-1} for 60 wt% PEG. The use of graphitic carbon nitride as shape-stabilisers reduces the extent of supercooling, while lowering the phase change enthalpy.

1.3 Metallic Nanoparticles

Metallic nanoparticles based on aluminium, copper and silver are used to enhance the thermal conductivity of composite PCM. Using PEG as PCM as well as reducing agent, in-situ reduction of AgNO$_3$ by PEG results in the formation of PEG/silver nanoparticles composite using a simple blending and heating process [19]. The thermal conductivity improves by 212% as compared to pure PEG with latent heat in the range of 121.9–141.9 J.g^{-1}.

Raw diatomite can be leached with alkali to dredge pores into diatomite, thereby improving the surface area and pore size (Figure 1.4a and b) [20]. Spherical, crystalline silver nanoparticles (Ag-NP) with particle size in the range of 3-10 nm are then uniformly distributed on the surface of diatomite

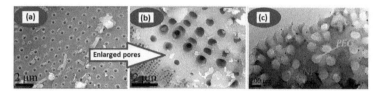

Figure 1.4 SEM micrographs of diatomite (a) after acid treatment, (b) alkali treatment to enlarge pores and (c) Silver decorated diatomite/PEG PCM. Reproduced from Ref. [6] by permission of The Royal Society of Chemistry.

and this Ag decorated diatomite is blended with PEG to enhance the heat transfer properties of PCM (Figure 1.4c). About 63 wt% PEG loading can lead to melting enthalpy of 111 J.g^{-1} at 60°C and reduced supercooling extent than that of PEG due to a heat conduction pathway provided by porous Ag-NP decorated diatomite network. With 7.2 wt% of Ag, a thermal conductivity of 0.82 W.m^{-1}.K^{-1} is possible, which is 127% higher than the system without it.

Aluminium nitride is added during in-situ polymerisation of methyl methacrylate to prepare PEG/polymethyl methacrylate (PMMA)/aluminium nitride composite withstanding maximum PEG encapsulation of 70 wt% [21]. By varying the mass fractions of aluminium nitride in the range 5-30%, the thermal conductivity can be increased from 0.253 to 0.389 W.m^{-1}.K^{-1}. However, the increasing mass fractions of aluminium nitride results in decrease of latent enthalpy from 116 to 79 J.g^{-1}. The volume resistivity increases from 0.26×10^{10} Ω.cm for pure PEG/PMMA PCM to 5.92×10^{10} Ω.cm for PEG/PMMA composite with 30 wt% aluminium nitride promoting electric insulation capabilities of the PCM.

To achieve the enhancement of thermal conductivity of PEG/SiO$_2$ composites, certain metal and carbon based fillers are used. The addition of β-aluminium nitride from 5% to 30 wt% enhances the thermal conductivity from 0.3847 to 0.7661 W m^{-1} K^{-1} [22]. Increase of the β-aluminium nitride mass fractions has adverse effect on enthalpy values as it reduces from 161 to 130 J.g^{-1}. Copper can be introduced in PEG/SiO$_2$ composites in-situ via chemical reduction of copper sulphate by applying ultrasound-assisted sol-gel method.[23] With 2.1 wt% copper, the composite exhibits 38% increase of thermal conductivity to 0.414 W m^{-1} K^{-1} while the melting enthalpy is exhibited at 110 J g^{-1}.

To PEG/expanded vermiculite composite, silver nanowire is added as thermal conductivity enhancer [24]. Silver nanowire of length 5-20 μm with

50–100 nm diameter, prepared by solvothermal method and wrapped by PEG, exhibits well-infused pores of expanded vermiculite. A maximum of 66.1 wt% of PEG can be incorporated for maintaining shape-stabilisation. Increase in concentration of silver nanowire increases the thermal conductivity, while reducing the enthalpy. The concentration of filler needs to be tuned in order for the PCM to have satisfactory latent heat for application. For example, with high weight fraction of 19.3 wt% silver nanowire the thermal conductivity increases by 11.3 times to a value of 0.68 $W.m^{-1}.K^{-1}$ with a melting enthalpy of 99 $J.g^{-1}$. A theoretical method developed for calculation of thermal conductivity values of PEG-silver/expanded vermiculite composite is consistent with the experimental values having an uncertainty of $\pm3\%$. Expanded vermiculite acts as a nucleating agent in promoting the crystallisation of PEG chains and causes reduction of supercooling extent by $7°C$. Silicon carbide nanowires (3.28 wt%) addition as conductive fillers to vermiculite/PEG composite results in 8.8 times higher conductivity than PEG to a value of 0.53 $W.m^{-1}.K^{-1}$ and melting enthalpy of 65 $J.g^{-1}$ [25].

1.4 Carbon Nanocomposites

Carbon based supporting materials such as activated carbon (AC), ordered mesoporous carbon (CMK-5), carbon nanotubes have gained significance, not only as thermally conductive fillers but as containment matrices. Such kind of properties improve the thermal conductivity of the composite and further, the porous nature of these materials allows for maximum accommodation of working material and offer structural stability without compromising the transition enthalpy significantly. Physical blending and impregnation method has been most commonly applied for preparation of these composites. Due to the porous nature of these materials, there is a minimum threshold for accommodation of working material below which the soft segment gets well absorbed into the pores, thereby hindering the crystallisation and thermal behaviour. For the PEG-mesoporous active carbon form-stable PCM, this threshold limit is 30 wt% PEG [26]. The blends with PEG wt% below 50 are found to be mesoporous with PEG getting adsorbed into the pores of AC. However, the blends with higher PEG wt% varying from 50-70% are found to be non-porous showing aggregates of PEG exhibiting easier ability to undergo crystallisation. The phase change activation energy increases with increasing AC content and decreasing PEG content. Similarly for larger chain of polyethylene glycol hexadecyl ether (Brij58) working material, this threshold limit is 25 wt% PEG with porous activated carbon supporting framework [27]. The non-isothermal crystallisation behaviour in Brij58/AC composite accounts for increase in half-time and relative degree of crystallinity indicating confined crystallisation of Brij58 as a result of nanoporous activated carbon matrix. The pore structure of these porous materials is found to affect the phase change behaviour of composites [28]. The irregular pores of AC had pore diameter of 4 nm whereas the hexagonally ordered pores of CMK-5 had a pore diameter of 3.5 nm. The mesoporous pores of EG have pore diameters of 13 μm. The phase change enthalpies of porous carbon composites and crystallinity of PEG increase in the order of PEG-AC<PEG-CMK-5<PEG-EG. This originates from the larger pore volume of EG whereas the small pore diameter of AC and CMK-5 hinder the crystallisation of PEG. Thus, higher pore volumes and ordered pore geometries are essential for shape-stabilisation.

Although commercially available carbon materials are most preferred for making PCM composites, there are efforts for preparing designed porous carbon from different raw materials. By carrying out controlled carbonisation of metal organic frameworks at different temperatures, highly porous carbons with exceptionally high surface area and large pore volume up to 2551.0 $m^2.g^{-1}$ and 3.1 $cm^3.g^{-1}$ respectively can be realised, which suits well for using as supporting matrix in PEG-based shape-stabilised PCMs [29]. Metal organic frameworks can be heated at high temperatures leading to the migration and evaporation of zinc oxide particles to form a highly porous matrix. The nano-cavities with large mesopore volume afford adsorption up to 92.5 wt% PEG with high latent heat of 162 $J.g^{-1}$. The disordered tiny graphitic porous carbon with very high specific surface area leads to enhanced thermal conductivity. In fact, the thermal conductivity can be improved by 50% by this strategy as compared to pure PEG. Porous carbon is also prepared from potato by using freeze drying and heat treatment techniques consecutively, thereby utilising renewable plant resources/agricultural feedstock [30]. The porous carbons prepared by this technique exhibit an average pore diameter of 204.7 nm with a 73.4% porosity, which can be loaded with a maximum of 50 wt% of PEG. This composite has reasonable mechanical strength to retain its shape-stability and can prevent leakage of melted PEG. Melted PEG, having good wettability to carbon, has exudation stability above melting temperatures. In addition, the extent of supercooling for same composite is reduced by 7.6%.

Various studies have focussed on using carbon nanomaterials for enhancement of thermal conductivity such as carbon fibre, carbon nanospheres (CNS), carbon nanotubes (CNT), multiwall carbon nanotubes (MWCNT) and single-walled carbon nanotubes (SWCNT), graphene and more.

1.4.1 Carbon Fibre

Structural laminates can be made by combining an epoxy resin with paraffin as PCM, which is stabilised with CNTs, and reinforcing carbon fibres [31]. The stabilised paraffin maintains inherent ability to melt and crystallise in the laminates even after repeated thermal cycling. The melting enthalpy of the composite is proportional to the paraffin weight fraction reaching a maximum value of 47.4 J cm^{-3}. Moreover, the thermal conductivity of the laminates through thickness direction increases with the content of CNT-stabilised PCM. A sharp drop of storage modulus occurs at T_g as well as at T_m of paraffin. Similarly, two tan δ peaks corresponding to transition in storage modulus exist. Above characteristics impart good impact properties to the laminates. Mechanical energy absorption capability of the laminates is revealed by a sequence of drops and plateaus in the load-displacement curves. Phase change composite from carbon fibre and erythritol can be made by the percolation of carbon fibre into erythritol. This can be carried out by two methods namely, melt-dispersion and hot press [32]. Hot press method is more effective for the preparation of phase change composite with carbon fibre due to alignment of carbon fibre in a plane perpendicular to pressing stress which increases the thermal conductivity exponentially as compared to conventional melt-dispersion method. Also, high packing ratio in hot press method requires smaller volume fraction of filler.

1.4.2 Carbon Nanospheres (CNS)

Carbon nanospheres (CNS) can be dispersed in PCM like stearic acid in varying weight percentages with maximum percentage of 50 wt% CNS [33]. Although, the thermal conductivity increases with increasing concentration of CNS from 0.21-0.431 W.m^{-1}.K^{-1} the enthalpy is adversely affected as the corresponding values decreases from 153 to 92 J.g^{-1}.

1.4.3 Carbon Nanotubes (CNT)

Polymer matrix has been developed that displays both the good capability of storing and releasing thermal energy together with good mechanical properties that make it suitable to fabricate structural composites. This has become possible by combining CNT confined paraffin as PCM with a high-performance epoxy matrix. PW, blended with 10 wt% CNT incorporated in epoxy matrix, exhibits very high phase change enthalpy value of 219.1 J.g^{-1} with higher thermal stability than PW. [34]. CNT acts as stabilising agent

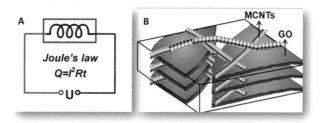

Figure 1.5 (a) TES by Joule heating and (b) schematic for GO/CNT network in PEG matrix. Reprinted (adapted) with permission from Ref. [35]. Copyright (2018) American Chemical Society.

allowing maximum incorporation of PW without leakage. For confining paraffin in CNTs, the latter is added to molten paraffin under mechanical stirring at 500 rpm for 30 min and cooled to ambient temperature by pouring in to silicon moulds. For shape stabilisation, minimum 10 wt% CNTs is required for containing paraffin. The CNT-confined paraffin, thus prepared can be added to epoxy resin and the mix is cured at ambient temperature followed by post curing at high temperature. Blending process does not affect the inherent melting and crystallisation peak temperatures. However, without CNT, the paraffin exudes out as the confinement action is missing in the blend. Paraffin loading in epoxy lowers the elastic modulus and flexural strength of the epoxy matrix with a trend in accordance with the mixture rule. The additional attribute of such materials is the decrease of electrical resistivity with loading of CNT.

Using CNTs in combination with graphene oxide (GO) to prepare PEG composites, a conductive interconnected pathway can be generated for self-heating by using low voltage as a result of Joule heating (Figure 1.5) [35]. The fusion enthalpy for 85 wt% PEG composite is 120.7 J.g^{-1} with storage capability of 66.3% of pure PEG. Joule heat storage efficiency of about 70% is achieved at 6.6 V. By crosslinking PEG and hexamethylene diisocyanate biuret to form polyurethane in presence of dispersed halloysite nanotubes (HNT), the fusion enthalpy improves from 86.6 to 118.7 J.g^{-1} with addition of just 1.02 wt% of HNT. [36]. In fact, HNT acts as crosslinking and nucleating agent so that the fusion enthalpy is improved.

Carbon nanotube sponge (CNTS) has been adopted as a porous scaffold to store PW by using solvent assisted infiltration to allow for uniform distribution of PW [37]. Enthalpy loadings beyond 91 wt% PW exceed the phase change enthalpy of pure PW but thermal conductivity saturation occurs at 80 wt% PW with six-fold increase compared to pure paraffin. The composite retains its shape stability without any deformation even under large compression. The extensive C-H...π interactions affect the storage efficiency of the composite. The composite is capable of storing energy under small voltages or light absorption maintaining thermal reliability for large number of cycles, e.g., beyond 100 cycles.

1.4.4 Multiwall Carbon Nanotubes (MWCNT)

Another carbon material, multiwall carbon nanotubes (MWCNT) and functionalised MWCNTs (-COOH, -NH$_2$, -OH functional groups) can be used to form nanocomposite with PEG [38]. PEG/functionalised MWCNTs have lower phase transition enthalpy and transition temperatures as compared to pure PEG/MWCNT PCM. The functional groups affect the enthalpy values of the composite PCMs in the order of MWCNT-COOH<MWCNT-NH$_2$<MWCNT-OH<MWCNT. The order is dictated by the combined influence of hydrogen bonding, capillary forces and surface adsorption. On dispersing MWCNT (3 wt%) in PEG/SiO$_2$ composite, thermal conductivity is improved to a value of 0.463 W.m^{-1}.K^{-1} [39]. The composite has multidirectional characteristics, such as rapid and broadband visible light-harvesting, light-thermal conversion together with thermal energy storage ability and form-stable effects. By the introduction of 2.7 wt% graphite to the PEG-SiO$_2$ composite the thermal conductivity can be enhanced to 0.558 W.m^{-1}.K^{-1} [40]. Paraffin/montmorillonite/grafted MWCNT composite, having latent heat of 47.7 J.g^{-1} and thermal conductivity of 0.301 W.m^{-1}.K^{-1}, can be prepared by using ultrasonic dispersion and liquid intercalation technique [41].

1.4.5 Single-walled Carbon Nanotubes (SWCNT)

Single-walled carbon nanotubes (SWCNT) with a high aspect ratio can be modified to incorporate nitrophenyl groups to allow dispersion in solvent, toluene and composite formation with PEG is possible due to π-π interactions between PCM and modified-SWCNT [42]. Though a maximum enthalpy of 103 J.g^{-1} can be achieved, the thermal conductivity is higher than the other carbon composites with MWCNT and carbon black. SWCNTs effectively capture photons and act as molecular heater by converting light into heat energy. With a small loading of 2 wt% of SWCNT, the thermal conductivity of the PEG/diatomite composite can be improved from 0.24

to 0.87 W.m^{-1}.K^{-1} without affecting the thermal energy storage properties [43]. About 60 wt% PEG can be incorporated in the composite with good thermal reliability for 200 thermal cycles. Enhancement of thermal conductivity by 250% is achieved with a minimum loading of 0.25% of SWCNT in n-octadecane [44]. This enhancement depends on several factors including type of CNT, functionalisation and purity. The enhancement effect of SWCNT in the composite is superior than the corresponding composite with graphene nanoplatelets (GNP), which is due to the possible alignment of alkane along the axis of CNT as compared to multiple orientations on GNP.

1.5 Graphene Nanocomposites

The research on graphite and its derivatives has intensified due to its two-dimensional nature, introducing high thermal conductivity and superior electronic and mechanical properties along with the added advantage of low density. Similar to carbon composites, these composites of graphene are also commonly prepared by using blending and impregnation process. Three-dimensional netlike architecture of graphene/PEG composites is able to contain 93 wt% PEG with fusion enthalpies reaching 166 J.g^{-1} [45]. This interconnected architecture of composite is used as a source of heat for thermoelectric device fabrication allowing long steady-state output time with increasing weight percentage of PEG. Another environment friendly derivative of graphene, sulphonated graphene (SG) is used as a nanocomposite matrix to incorporate 96 wt% PEG by using solution processing in aqueous medium [46]. A four-fold increase in thermal conductivity from 0.263 W.m^{-1}.K^{-1} for pristine PEG to 1.042 W.m^{-1}.K^{-1} for the composite is possible, while suffering a 12.9% reduction in latent heat value.

1.5.1 Graphene Oxide (GO) and Derivatives

Many oxygen functional groups of graphene oxide (GO) undergo strong hydrogen bonding interactions with PEG, which facilitates about 90 wt% incorporation of PEG [47]. GO is effective in lowering phase change temperature without much impact on phase change enthalpy as compared to other carbon based porous materials including activated carbon and CMK-5.

By using both blending-impregnation and microwave method, 96 wt% PEG/GO nanocomposites can be prepared [48,49]. The use of microwave enhanced interlayer spacing between GO sheets for impregnation allows efficient dispersion of the PEG molecular chains. This results in high TES

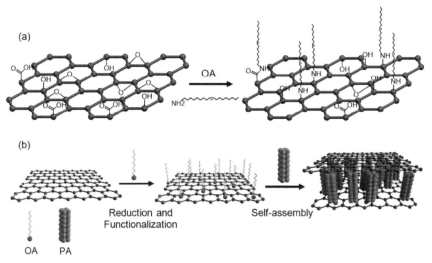

Figure 1.6 (a) Functionalisation of GO. (b) Reduction, functionalisation and self-assembly process. Reprinted (adapted) with permission from Ref. [51]. Copyright (2015) American Chemical Society.

of 175 J.g^{-1} for the composite prepared by microwave method. However, the composite, prepared by blending and impregnation method, has lower latent heat of 143 J.g^{-1}. Similar to other GO composites, this composite also exhibits improved photo to thermal energy conversion with increase in concentration of GO. The surface modification of GO can be done to make reduced graphene oxide (rGO) and GO-COOH respectively [50]. The melting points and phase change enthalpies decrease in the order of PEG/GO-COOH < PEG/GO < PEG/rGO. Maximum depression of melting point is exhibited by PEG/GO-COOH by 16.8°C as compared to PEG. This can help to manipulate the behaviour by proper functionalisation of the surface. Long chain alkyl amine, oleylamine are used to functionalise reduced graphene oxide (rGO) and palmitic acid is adsorbed onto the network in varying percentages (Figure 1.6) [51]. With a minimum loading of 0.6 wt% of modified GO, high phase change enthalpy of 196.6 J.g^{-1} and thermal conductivity enhancement by 150% as compared to pure palmitic acid can be achieved. The photo to heat conversion efficiency is quite high and the thermal reliability also remains high as (up to 500 thermal cycles) GO sheets are impregnated with 71 wt% stearic acid (SA) to function as PCM, which exhibits phase change enthalpy of 56 J.g^{-1} [52]. SA is confined in the interlayer spaces of the composite having thermal storage capabilities of 82.4%.

1.5.2 Graphene Aerogels (GA)

Three-dimensional hybrid graphene aerogels (HGA) can act as both nucleating agent and supporting material allowing incorporation of 2 wt% GO and 4 wt% GNP by using vacuum impregnation and physical blending methods (Figure 1.6) [53,54]. These composites exhibit high latent heat values in the range of 170-186 J.g^{-1}. In addition, the three-dimensional supporting framework leads to significant increase in thermal conductivity. The vacuum impregnated composite exhibits an increase of thermal conductivity from 0.3 W.m^{-1}.K^{-1} of pure PEG to a value of 1.43 W.m^{-1}.K^{-1} for the composition of 0.45 wt% GO and 1.8 wt% GNP. Additionally, the PCM composite with lowest filler content of 0.5 wt% HGA shows light to thermal energy conversion with 92% efficiency and the efficiency reduced with increase in filler content. For the composite prepared by using physical blending method, the thermal conductivity reaches a maximum of 1.7 W.m^{-1}.K^{-1} and electrical conductivity up to 2.5 S.m^{-1}. In presence of ultralow filler concentrations, this composite exhibits high latent heat values of 178 J.g^{-1} which is 98.2% of that of pure PEG. Using modified hydrothermal method, GO sheets are reduced into aerogels and used as shell to encapsulate core, paraffin [55]. A very high encapsulation of 97 wt% paraffin is achieved with high phase change enthalpy of 202 J.g^{-1}. The thermal conductivity increases by 32% with respect to pure paraffin.

1.5.3 Expanded Graphite (EG)

The porous structure of expanded graphite (EG) is encapsulated with paraffin, n-docosane, having maximum of 10 wt% EG [56]. With thermal conductivity of 0.82 W.m^{-1}.K^{-1}, the melting times reduce with increasing concentrations of EG. Also, the enthalpy for 10 wt% EG composite is 178.3 J.g^{-1}, lower by 8.4% as compared to pure paraffin. The thermal conductivity increases by 10 folds to a value of 3.83 W.m^{-1}.K^{-1} for 10 wt% EG/paraffin composite [57]. To improve the compatibility between paraffin and hydroxyl-terminated polybutadiene (HTPB), form-stable composite is prepared by physical blending with nano-silica as support and EG as thermally conductive fillers (Figure 1.7) [58]. The incorporation of the PW/nano-SiO$_2$/EG composite in the HTPB matrix reduces the thermal energy storage capacity significantly due to the hindrance posed to crystallisation. The addition of EG drastically improves the thermal conductivity from 0.303 to 0.602 W.m^{-1}.K^{-1} as compared to pure PW. Blends of PEG and expanded graphite (EG), containing 90 wt% PEG, have high enthalpy of 161 J.g^{-1} at 61°C

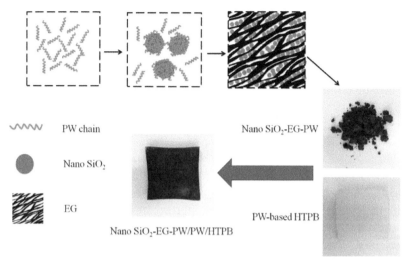

Figure 1.7 Schematic for preparation of nano-silica/EG/PW/HTPB composite. Reprinted (adapted) with permission from Ref. [58]. Copyright (2018) American Chemical Society.

[59]. Further, the thermal conductivity improves with increase in percentage incorporation of EG with maximum values reaching 1.324 W.m^{-1}.K^{-1}. EG prepared using microwave method is incorporated with n-octadecane to prepare phase change composite with 90 wt% soft segment possessing phase change enthalpy of 187 J.g^{-1} [60]. The thermal diffusivity of the composite increases with EG content which increases by 296.3% as compared to pure octadecane. The same EG is also blended with fatty alcohol, tetradecanol using autoclave method in the presence of solvent, ethanol [61]. 93 wt% of tetradecanol can be incorporated with high enthalpy of 202.6 J.g^{-1}.

To target medium temperature applications, sebacic acid/EG composite can be made by dry pressing, which has a phase change enthalpy of 187 J.g^{-1} at 128°C. Negligible subcooling and high thermal conductivity of 5.353 W.m^{-1}.K^{-1} are the two important properties of the nanocomposite [62]. Ternary mixture of myristic, palmitic and stearic acid can be blended with EG incorporating 93 wt% of the eutectic mixture [63]. Maintaining improved thermal conductivity of 2.51 W.m^{-1}.K^{-1}, the composite exhibits good fusion and crystallisation enthalpy of 153.5 and 151.4 J.g^{-1} respectively.

1.5.4 Graphene Nanoplatelets (GNPs)

One of the alkanes, eicosane, is used to prepare composite PCM with graphene nanoplatelets (GNP) under sonication and in the absence of surfactant [64]. With a maximum loading of 10 wt% GNP, thermal conductivity can be enhanced by 400% from that of pure eicosane. Furthermore, the thermal conductivity increases significantly with increase in temperature with optimum range being 30–35°C for this composite. Consequently, the thermal interface resistance reduces due to two dimensional planar structure which aids high performance of GNP composite. The reduction of thermal energy storage density by 16% for 10 wt% GNP, still keeps the latent enthalpy considerably high at a value of 220 $J.g^{-1}$.

GNPs serving as the conductive fillers and polyethylene glycol (PEG) acting as the phase change material (PCM) can be uniformly dispersed and embedded inside the network structure of polymethyl methacrylate (PMMA), which imparts well package and self-supporting properties of composite [65]. Followed by the sonication of expanded graphite to obtain GNP, it is immersed into the polymer matrix of PMMA in the presence of 70 wt% PEG. Incorporating a mass fraction of 8 wt% GNP, the thermal conductivity can be enhanced by 9 times as compared to pure PEG. This addition of GNP also improves the electrical conductivity by 8 orders with improvement from 10^{-9} $S.cm^{-1}$ for 1 wt% GnP to 10^{-4} $S.cm^{-1}$ for 2 wt% GnP. However, the latent heat varies from 125 to 114 $J.g^{-1}$ with increasing GNP content. GNP aids in reducing the supercooling extent and enhances the thermal stability of the matrix making it feasible for use in applications such as EMI shielding, anti-electrostatic material and bipolar plates in proton exchange membrane fuel cells.

To improve the thermal conductivity of composite PCM, exfoliated GNP can be vacuum impregnated with bio-based PCMs obtained from under-used feedstocks of fatty acid esters [66]. Although the thermal conductivity improves from 0.154 to 0.557 $W.m^{-1}.K^{-1}$, the latent heat storage capacity reduces to large extent (25%) from that of the pure bio-based PCM.

PEG is vacuum impregnated into three-dimensional porous cellulose/GNPs aerogels to fabricate phase change composites [67]. The high thermal conductivity of GNPs increases the heat carrying capacity to large extent with a value of 1.35 $W.m^{-1}.K^{-1}$ at 5.3 wt% GNP. Fusion enthalpy of 87% of that of PEG is retained by the composite with a value of 156 $J.g^{-1}$ at 89 wt% PEG. Polyurethane-GNP composites have been prepared from butane diol with or without chain extender [68]. The concentrations of GNP are maintained below 1% for nucleation efficiency and higher degree of crystallinity. About 0.3 wt% is maintained for composites without chain extender and 0.5 wt% for composites with chain extender. The enthalpies of composites are higher without chain extenders ranging between 118.0–164.5

J.g^{-1} as compared to enthalpies of composites with chain extender ranging between 128.0-148.5 J.g^{-1}. The two-dimensional pathway provided by GNP for phonon transport increases the thermal conductivity of both the composites significantly. Further, the ultrasonic wave propagation increases with increasing GNP content which indicates better crystals formation [69].

1.5.5 Graphite Foam (GF)

By carrying out in-situ polymerisation of PEG and diisocyanates within graphite foam (GF), smart PCM composite with electro/photo to heat conversion is developed with more than 80 wt% of polyurethane (PU) in the composite [70]. The thermal conductivity enhances by 12 times as compared to PEG, to a value of 3.5 W.m^{-1}.K^{-1}. However, the enthalpies are lower at a value of 80.3 J.g^{-1} due to urethane linkages restricting the movement of PEG chain and reducing the hydroxyl groups of PEG inhibiting intermolecular hydrogen bonding. The composite shows storage efficiency of more than 80% for electro to heat conversion and 67% efficiency for photo to heat conversion.

Macroporous pitch-based GF is loaded with 73 wt% in situ cross-linked polyurethane yielding enthalpy of 60 J.g^{-1} [71]. However, supercooling is significantly reduced and thermal conductivity improves by 43 times as compared to PU with a high value of 10.86 W.m^{-1}.K^{-1}. With the thermal storage efficiency reaching up to 85%, the composite is successful in providing thermal buffering as a wear layer for protection from cold.

The filling of GF with interconnected hollow graphene network results in the formation of hierarchical GF (HGF) and reduces the thermal resistance by providing a denser three-dimensional conduction pathway [72]. PW is immersed in HGF pores by vacuum impregnation method which enhances the thermal conductivity by 744% as compared to pure PW. Further, the latent enthalpy is not much affected remaining at 95.2% of PW having good reliability up to 100 thermal cycles. This composite is also effective in photoabsorption with 89% thermal energy storage efficiency.

1.6 Conclusions

A wide range of nanocomposite PCMs have been explored as solid-solid PCMs. The varieties of materials presented have highly porous nature which can accommodate high percentage of PCM, high thermal energy storage density without liquid leakage and high thermal conductivity. In this chapter, we explored a wide array of nanomaterials used in conjugation with organic

PCMs. Three major categories of nanomaterials include metallic, carbon-based and graphene-based nanomaterials. The use of nanomaterials enhances the thermal conductivity but on the other hand, it leads to significant decline of thermal energy storage efficiency. Among the various materials explored, graphene foam and single-walled carbon nanotubes offers huge enhancement of thermal conductivity resulting in the preparation of smart PCMs with efficient photo to thermal conversion without sacrificing the thermal energy storage capacity. For the commercial success of these materials, right balance of working material and thermally conductive filler/supporting matrix needs to be established depending on the end application.

References

[1] Faninger G. Thermal energy storage. In: Faculty for Interdisciplinary Research and Continuing Education. Austria: IFF-University of Klagenfurt, 2004.

[2] Kenisarin MM, Kenisarina KM. Form-stable phase change materials for thermal energy storage. Renew Sustain Energy Rev 2012;16:1999–2040. doi:10.1016/j.rser.2012.01.015.

[3] Sharma A, Tyagi VV, Chen CR, Buddhi D. Review on thermal energy storage with phase change materials and applications. Renew Sustain Energy Rev 2009;13:318–45. doi:10.1016/j.rser.2007.10.005.

[4] Mehling H, Cabeza LF. Solid-liquid phase change materials. In: Heat and Cold Storage with PCM: An Up to Date Introduction into Basics and Applications, 2008, pp. 11–55.

[5] Mondal S. Phase change materials for smart textiles - An overview. Appl Therm Eng 2008;28:1536–50. doi:10.1016/j.applthermaleng.2007.08.009.

[6] Sundararajan S, Samui AB, Kulkarni PS. Versatility of polyethylene glycol (PEG) in designing solid-solid phase change materials (PCMs) for thermal management and their application to innovative technologies. J Mater Chem A 2017;5. doi:10.1039/c7ta04968d.

[7] Hyun DC, Levinson NS, Jeong U, Xia Y. Emerging applications of phase-change materials (PCMs): Teaching an old dog new tricks. Angew Chemie Int Ed 2014;53:3780–95. doi:10.1002/anie.201305201.

[8] Lv J, Song Y, Jiang L, Wang J. Bio-inspired strategies for anti-icing. ACS Nano 2014;8:3152–69. doi:10.1021/nn406522n.

[9] Sharma RK, Ganesan P, Tyagi VV, Metselaar HSC, Sandaran SC. Developments in organic solid–liquid phase change materials and

their applications in thermal energy storage. Energy Convers Manag 2015;95:193–228. doi:10.1016/j.enconman.2015.01.084.

[10] Pielichowska K, Pielichowski K. Phase change materials for thermal energy storage. Prog Mater Sci 2014;65:67–123. doi:10.1016/j.pmatsci. 2014.03.005.

[11] Sundararajan S, Samui AB, Kulkarni PS. Shape-stabilized poly(ethylene glycol) (PEG)-cellulose acetate blend preparation with superior PEG loading via microwave-assisted blending. Sol Energy 2017;144. doi:10.1016/j.solener.2016.12.056.

[12] Min X, Fang M, Huang Z, Liu Y, Huang Y, Wen R, et al. Enhanced thermal properties of novel shape-stabilized PEG composite phase change materials with radial mesoporous silica sphere for thermal energy storage. Sci Rep 2015;5:12964. doi:10.1038/srep12964.

[13] Kant K, Shukla A, Sharma A. Advancement in phase change materials for thermal energy storage applications. Sol Energy Mater Sol Cells 2017;172:82–92. doi:10.1016/j.solmat.2017.07.023.

[14] Fang X, Fan L-W, Ding Q, Yao X-L, Wu Y-Y, Hou J-F, et al. Thermal energy storage performance of paraffin-based composite phase change materials filled with hexagonal boron nitride nanosheets. Energy Convers Manag 2014;80:103–9. doi:10.1016/j.enconman.2014.01.016.

[15] Li Y, Li J, Feng W, Wang X, Nian H. Design and preparation of the phase change materials paraffin/porous Al_2O_3 @Graphite foams with enhanced heat storage capacity and thermal conductivity. ACS Sustain Chem Eng 2017;5:7594–603. doi:10.1021/acssuschemeng.7b00889.

[16] Yang J, Tang LS, Bao RY, Bai L, Liu ZY, Xie BH, et al. Hybrid network structure of boron nitride and graphene oxide in shape-stabilized composite phase change materials with enhanced thermal conductivity and light-to-electric energy conversion capability. Sol Energy Mater Sol Cells 2018;174:56–64. doi:10.1016/j.solmat.2017.08.025.

[17] Deng Y, Li J, Nian H, Li Y, Yin X. Design and preparation of shape-stabilized composite phase change material with high thermal reliability via encapsulating polyethylene glycol into flower-like TiO_2 nanostructure for thermal energy storage. Appl Therm Eng 2017;114:328–36. doi:10.1016/j.applthermaleng.2016.11.082.

[18] Feng L, Song P, Yan S, Wang H, Wang J. The shape-stabilized phase change materials composed of polyethylene glycol and graphitic carbon nitride matrices. Thermochim Acta 2015;612:19–24. doi:10.1016/j.tca. 2015.05.001.

[19] Liang W, Wang L, Zhu Z, Qian C, Sun H, Yang B, et al. In situ preparation of polyethylene glycol/silver nanoparticles composite phase change materials with enhanced thermal conductivity. ChemistrySelect 2017;2:3428–36. doi:10.1002/slct.201700381.

[20] Qian T, Li J, Min X, Guan W, Deng Y, Ning L. Enhanced thermal conductivity of PEG/diatomite shape-stabilized phase change materials with Ag nanoparticles for thermal energy storage. J Mater Chem A 2015;3:8526–36. doi:10.1039/C5TA00309A.

[21] Zhang L, Zhu J, Zhou W, Wang J, Wang Y. Characterization of polymethyl methacrylate/polyethylene glycol/aluminum nitride composite as form-stable phase change material prepared by in situ polymerization method. Thermochim Acta 2011;524:128–34. doi:10.1016/j.tca.2011.07.003.

[22] Wang W, Yang X, Fang Y, Ding J, Yan J. Enhanced thermal conductivity and thermal performance of form-stable composite phase change materials by using β-aluminum nitride. Appl Energy 2009;86:1196–200. doi:10.1016/j.apenergy.2008.10.020.

[23] Tang B, Qiu M, Zhang S. Thermal conductivity enhancement of PEG/SiO2 composite PCM by in situ Cu doping. Sol Energy Mater Sol Cells 2012;105:242–8. doi:10.1016/j.solmat.2012.06.012.

[24] Deng Y, Li J, Qian T, Guan W, Li Y, Yin X. Thermal conductivity enhancement of polyethylene glycol/expanded vermiculite shape-stabilized composite phase change materials with silver nanowire for thermal energy storage. Chem Eng J 2016;295:427–35. doi:10.1016/j.cej.2016.03.068.

[25] Deng Y, Li J, Nian H. Polyethylene glycol-enwrapped silicon carbide nanowires network/expanded vermiculite composite phase change materials: Form-stabilization, thermal energy storage behavior and thermal conductivity enhancement. Sol Energy Mater Sol Cells 2018;174:283–91. doi:10.1016/j.solmat.2017.09.013.

[26] Feng L, Zheng J, Yang H, Guo Y, Li W, Li X. Preparation and characterization of polyethylene glycol/active carbon composites as shape-stabilized phase change materials. Sol Energy Mater Sol Cells 2011;95:644–50. doi:10.1016/j.solmat.2010.09.033.

[27] Zhang L, Shi H, Li W, Han X, Zhang X. Thermal performance and crystallization behavior of poly(ethylene glycol) hexadecyl ether in confined environment. Polym Int 2014;63:982–8. doi:10.1002/pi.4592.

[28] Wang C, Feng L, Li W, Zheng J, Tian W, Li X. Shape-stabilized phase change materials based on polyethylene glycol/porous carbon composite: The influence of the pore structure of the carbon materials. Sol Energy Mater Sol Cells 2012;105:21–6. doi:10.1016/j.solmat.2012.05.031.

[29] Tang J, Yang M, Dong W, Yang M, Zhang H, Fan S, et al. Highly porous carbons derived from MOFs for shape-stabilized phase change materials with high storage capacity and thermal conductivity. RSC Adv 2016;6:40106–14. doi:10.1039/C6RA04059D.

[30] Tan B, Huang Z, Yin Z, Min X, Liu Y, Wu X, et al. Preparation and thermal properties of shape-stabilized composite phase change materials based on polyethylene glycol and porous carbon prepared from potato. RSC Adv 2016;6:15821–30. doi:10.1039/C5RA25685B.

[31] Fredi G, Dorigato A, Fambri L, Pegoretti A. Multifunctional epoxy/carbon fiber laminates for thermal energy storage and release. Compos Sci Technol 2018;158:101–11. doi:10.1016/j.compscitech.2018.02.005.

[32] Nomura T, Tabuchi K, Zhu C, Sheng N, Wang S, Akiyama T. High thermal conductivity phase change composite with percolating carbon fiber network. Appl Energy 2015;154:678–85. doi:10.1016/j.apenergy.2015.05.042.

[33] Mehrali M, Tahan Latibari S, Mehrali M, Mahlia TMI, Cornelis Metselaar HS. Effect of carbon nanospheres on shape stabilization and thermal behavior of phase change materials for thermal energy storage. Energy Convers Manag 2014;88:206–13. doi:10.1016/j.enconman.2014.08.014.

[34] Fredi G, Dorigato A, Fambri L, Pegoretti A. Wax confinement with carbon nanotubes for phase changing epoxy blends. Polymers 2017;9:405. doi:10.3390/polym9090405.

[35] Guo X, Liu C, Li N, Zhang S, Wang Z. Electrothermal conversion phase change composites: The case of polyethylene glycol infiltrated graphene oxide/carbon nanotube networks. Ind Eng Chem Res 2018:acs.iecr.8b03093. doi:10.1021/acs.iecr.8b03093.

[36] Zhou Y, Sheng D, Liu X, Lin C, Ji F, Dong L, et al. Synthesis and properties of crosslinking halloysite nanotubes/polyurethane-based solid-solid phase change materials. Sol Energy Mater Sol Cells 2018;174:84–93. doi:10.1016/j.solmat.2017.08.031.

[37] Chen L, Zou R, Xia W, Liu Z, Shang Y, Zhu J, et al. Electro- and photodriven phase change composites based on wax-infiltrated carbon nanotube sponges. ACS Nano 2012;6:10884–92. doi:10.1021/nn304310n.

[38] Feng L, Wang C, Song P, Wang H, Zhang X. The form-stable phase change materials based on polyethylene glycol and functionalized carbon nanotubes for heat storage. Appl Therm Eng 2015;90:952–6. doi:10.1016/j.applthermaleng.2015.07.080.

[39] Tang B, Wang Y, Qiu M, Zhang S. A full-band sunlight-driven carbon nanotube/PEG/SiO2 composites for solar energy storage. Sol Energy Mater Sol Cells 2014;123:7–12. doi:10.1016/j.solmat.2013.12.022.

[40] Li J, He L, Liu T, Cao X, Zhu H. Preparation and characterization of PEG/SiO2 composites as shape-stabilized phase change materials for thermal energy storage. Sol Energy Mater Sol Cells 2013;118:48–53. doi:10.1016/j.solmat.2013.07.017.

[41] Li M, Guo Q, Nutt S. Carbon nanotube/paraffin/montmorillonite composite phase change material for thermal energy storage. Sol Energy 2017;146:1–7. doi:10.1016/j.solener.2017.02.003.

[42] Wang Y, Tang B, Zhang S. Single-walled carbon nanotube/phase change material composites: Sunlight-driven, reversible, form-stable phase transitions for solar thermal energy storage. Adv Funct Mater 2013;23:4354–60. doi:10.1002/adfm.201203728.

[43] Qian T, Li J, Feng W, Nian H. Enhanced thermal conductivity of form-stable phase change composite with single-walled carbon nanotubes for thermal energy storage. Sci Rep 2017;7:44710. doi:10.1038/srep44710.

[44] Harish S, Ishikawa K, Chiashi S, Shiomi J, Maruyama S. Anomalous thermal conduction characteristics of phase change composites with single-walled carbon nanotube inclusions. J Phys Chem C 2013;117:15409–13. doi:10.1021/jp4046512.

[45] Jiang Y, Wang Z, Shang M, Zhang Z, Zhang S. Heat collection and supply of interconnected netlike graphene/polyethyleneglycol composites for thermoelectric devices. Nanoscale 2015;7:10950–3. doi:10.1039/C5NR02051D.

[46] Li H, Jiang M, Li Q, Li D, Chen Z, Hu W, et al. Aqueous preparation of polyethylene glycol/sulfonated graphene phase change composite with enhanced thermal performance. Energy Convers Manag 2013;75:482–7. doi:10.1016/j.enconman.2013.07.005.

[47] Wang C, Feng L, Yang H, Xin G, Li W, Zheng J, et al. Graphene oxide stabilized polyethylene glycol for heat storage. Phys Chem Chem Phys 2012;14:13233–8. doi:10.1039/c2cp41988b.

[48] Qi G-Q, Liang C-L, Bao R-Y, Liu Z-Y, Yang W, Xie B-H, et al. Polyethylene glycol based shape-stabilized phase change material for thermal

energy storage with ultra-low content of graphene oxide. Sol Energy Mater Sol Cells 2014;123:171–7. doi:10.1016/j.solmat.2014.01.024.

[49] Xiong W, Chen Y, Hao M, Zhang L, Mei T, Wang J, et al. Facile synthesis of PEG based shape-stabilized phase change materials and their photo-thermal energy conversion. Appl Therm Eng 2015;91:630–7. doi:10.1016/j.applthermaleng.2015.08.063.

[50] Wang C, Wang W, Xin G, Li G, Zheng J, Tian W, et al. Phase change behaviors of PEG on modified graphene oxide mediated by surface functional groups. Eur Polym J 2016;74:43–50. doi:10.1016/j.eurpolymj.2015.10.027.

[51] Akhiani AR, Mehrali M, Tahan Latibari S, Mehrali M, Mahlia TMI, Sadeghinezhad E, et al. One-step preparation of form-stable phase change material through self-assembly of fatty acid and graphene. J Phys Chem C 2015;119:22787–96. doi:10.1021/acs.jpcc.5b06089.

[52] Li B, Liu T, Hu L, Wang Y, Nie S. Facile preparation and adjustable thermal property of stearic acid–graphene oxide composite as shape-stabilized phase change material. Chem Eng J 2013;215–216:819–26. doi:10.1016/j.cej.2012.11.077.

[53] Yang J, Qi G, Liu Y, Bao R, Liu Z, Yang W, et al. Hybrid graphene aerogels/phase change material composites: Thermal conductivity, shape-stabilization and light-to-thermal energy storage. Carbon 2016;100:693–702. doi:10.1016/j.carbon.2016.01.063.

[54] Qi G-Q, Yang J, Bao R-Y, Liu Z-Y, Yang W, Xie B-H, et al. Enhanced comprehensive performance of polyethylene glycol based phase change material with hybrid graphene nanomaterials for thermal energy storage. Carbon 2015;88:196–205. doi:10.1016/j.carbon.2015.03.009.

[55] Ye S, Zhang Q, Hu D, Feng J. Core–shell-like structured graphene aerogel encapsulating paraffin: shape-stable phase change material for thermal energy storage. J Mater Chem A 2015;3:4018–25. doi:10.1039/C4TA05448B.

[56] Sarí A, Karaipekli A. Thermal conductivity and latent heat thermal energy storage characteristics of paraffin/expanded graphite composite as phase change material. Appl Therm Eng 2007;27:1271–7. doi:10.1016/j.applthermaleng.2006.11.004.

[57] Xia L, Zhang P, Wang RZ. Preparation and thermal characterization of expanded graphite/paraffin composite phase change material. Carbon 2010;48:2538–48. doi:10.1016/j.carbon.2010.03.030.

[58] Gao X, Zhao T, Luo G, Zheng B, Huang H, Han X, et al. Thermal property enhancement of paraffin-wax-based hydroxyl-terminated polybutadiene binder with a novel nanoSiO$_2$-expanded graphite-PW ternary form-stable phase change material. Energy & Fuels 2018;32:4016–24. doi:10.1021/acs.energyfuels.7b03856.

[59] Wang W, Yang X, Fang Y, Ding J, Yan J. Preparation and thermal properties of polyethylene glycol/expanded graphite blends for energy storage. Appl Energy 2009;86:1479–83. doi:10.1016/j.apenergy.2008.12.004.

[60] Li H, Liu X, Fang G-Y. Synthesis and characteristics of form-stable n-octadecane/expanded graphite composite phase change materials. Appl Phys A 2010;100:1143–8. doi:10.1007/s00339-010-5724-y.

[61] Zeng J-L, Gan J, Zhu F-R, Yu S-B, Xiao Z-L, Yan W-P, et al. Tetradecanol/expanded graphite composite form-stable phase change material for thermal energy storage. Sol Energy Mater Sol Cells 2014;127:122–8. doi:10.1016/j.solmat.2014.04.015.

[62] Wang S, Qin P, Fang X, Zhang Z, Wang S, Liu X. A novel sebacic acid/expanded graphite composite phase change material for solar thermal medium-temperature applications. Sol Energy 2014;99:283–90. doi:10.1016/j.solener.2013.11.018.

[63] Yang X, Yuan Y, Zhang N, Cao X, Liu C. Preparation and properties of myristic–palmitic–stearic acid/expanded graphite composites as phase change materials for energy storage. Sol Energy 2014;99:259–66. doi:10.1016/j.solener.2013.11.021.

[64] Fang X, Fan L-W, Ding Q, Wang X, Yao X-L, Hou J-F, et al. Increased thermal conductivity of eicosane-based composite phase change materials in the presence of graphene nanoplatelets. Energy & Fuels 2013;27:4041–7. doi:10.1021/ef400702a.

[65] Zhang L, Zhu J, Zhou W, Wang J, Wang Y. Thermal and electrical conductivity enhancement of graphite nanoplatelets on form-stable polyethylene glycol/polymethyl methacrylate composite phase change materials. Energy 2012;39:294–302. doi:10.1016/j.energy.2012.01.011.

[66] Jeong SG, Chung O, Yu S, Kim S, Kim S. Improvement of the thermal properties of bio-based PCM using exfoliated graphite nanoplatelets. Sol Energy Mater Sol Cells 2013;117:87–92. doi:10.1016/j.solmat.2013.05.038.

[67] Yang J, Zhang E, Li X, Zhang Y, Qu J, Yu Z. Cellulose/graphene aerogel supported phase change composites with high thermal conductivity and good shape stability for thermal energy storage. Carbon N Y 2016;98:50–7. doi:10.1016/j.carbon.2015.10.082.

[68] Pielichowska K, Bieda J, Szatkowski P. Polyurethane/graphite nano-platelet composites for thermal energy storage. Renew Energy 2016;91:456–65. doi:10.1016/j.renene.2016.01.076.

[69] Pielichowska K, Nowak M, Szatkowski P, Macherzyńska B. The influ-ence of chain extender on properties of polyurethane-based phase change materials modified with graphene. Appl Energy 2016;162:1024–33. doi:10.1016/j.apenergy.2015.10.174.

[70] Chen R, Yao R, Xia W, Zou R. Electro/photo to heat conversion system based on polyurethane embedded graphite foam. Appl Energy 2015;152:183–8. doi:10.1016/j.apenergy.2015.01.022.

[71] Wu W, Huang X, Li K, Yao R, Chen R, Zou R. A func-tional form-stable phase change composite with high efficiency electro-to-thermal energy conversion. Appl Energy 2017;190:474–80. doi:10.1016/j.apenergy.2016.12.159.

[72] Qi G, Yang J, Bao R, Xia D, Cao M, Yang W, et al. Hierar-chical graphene foam-based phase change materials with enhanced thermal conductivity and shape stability for efficient solar-to-thermal energy conversion and storage. Nano Res 2017;10:802–13. doi:10.1007/s12274-016-1333-1.

2

Fabrication of Natural Dye-Sensitised Solar Cells Based on Quasi Solid State Electrolyte Using TiO$_2$ Nanocomposites

N. Suganya[1], G. Hari Hara Priya[2] and V. Jaisankar[1*]

[1]PG and Research Department of Chemistry, Presidency College (Autonomous), Chennai, India
[2]Department of Chemistry, SDNB Vaishnav College for Women, Chromepet, Chennai, India
E-mail: vjaisankar@gmail.com
*Corresponding Author

Natural dye-sensitised solar cells (NDSSCs) have emerged recently as efficient and low-cost photovoltaic cells. The natural dye molecules were used as the light harvesting material in NDSSCs. The core shell nanoparticles of solid-state absorber are used as a sensitiser in conjunction with the *Opuntia stricta* fruit dye. TiO$_2$-CuO and TiO$_2$-ZnO nanocomposites were synthesised using sol-gel method, which were used as photoanode. The liquid electrolyte was replaced by quasi-solid gel polymer electrolytes (GPEs) using poly(methyl methacrylate) and polyethylene glycol which are incorporated into the liquid iodine/iodide electrolyte matrix. The natural dye obtained from *Opuntia stricta* fruit was characterised using ultraviolet visible spectroscopy. The gel polymer electrolyte was characterised using FT-IR; scanning electron microscopy (SEM) and ionic conductivity were studied using electrochemical impedance spectroscopy (EIS). The current voltage measurements of solar cell based on *Opuntia stricta* fruit dye-sensitised TiO$_2$ and TiO$_2$/CuO films showed an enhancement of open circuit voltage (V_{oc}) from 0.43 to 0.49 V, when CuO is added. This confirms that CuO plays an active role for the energy barrier leading to decrease in recombination losses.

2.1 Introduction

Natural dye-sensitised solar cells (DSSCs) are a potential alternative to the traditional photovoltaic devices. It has gained considerable attention in the field of solar energy due to their simple fabrication, environmental friendly materials endowed with good efficiencies and low production cost [1]. DSSCs consist of nanoporous wide band gap semiconducting oxide layer, dye sensitiser which plays important role in harvesting sunlight and transforming solar energy into electric energy, an electrolyte and a counter electrode which may be graphite coated or platinum coated.

Besides TiO_2, semiconductors like SnO_2, Fe_2O_3, ZrO_2, Al_2O_3 and ZnO have also been well studied for solar cell applications. The recombination of photo-injected electrons in the conduction band of TiO_2 with the oxidised dye is one of the most important problems in TiO_2. The relatively slow electron transport rate resulting from multiple trapping/detrapping events occurring within grain boundaries, will lead to high interface recombination and limit the device efficiency. The best approach to solve the limitation exhibited by TiO_2 as photoanode is achieved by the replacement of core/shell nanostructures. Based on a hypothesis that a coating layer may build up an energy barrier at the semiconductor/electrolyte interface which in turn lowers the charge recombination. Thus, retard the reaction between the photogenerated electrons and the redox species in electrolyte. More works have been focused on the development of core–shell nanostructured photoanodes such as $TiO_2@ZnO$ [2-5], $TiO_2@Nb_2O_5$ [6], $TiO_2@SnO_2$ [7], $TiO_2@Al_2O_3$ [8], $TiO_2@ZrO_2$ [9] and $ZnO@Cu_2O$ [10]. It was reported that with these nanostructures, a favourable energy gradient in the bulk of the anode can be developed to enhance electron transport.

Next to photoanode, electrolyte plays an important role in DSSC. Recently, many groups have been working on polymer electrolyte to improve transport properties. Another crucial parameter in the fabrication of DSSCs is the sensitising dye. The alternative for organic dye is natural dye. It is environmental friendly and low in cost.

In this present work, we discussed the preparation of TiO_2, TiO_2-ZnO, TiO_2-CuO and polymer gel electrolyte. The effects of these oxide layers on the performance of DSSCs assembled with a gel polymer electrolyte were investigated. The prepared core/shell nanostructures and gel polymer electrolytes were characterised by various analytical methods. The natural dye obtained from *Opuntia stricta* fruit was characterised using Ultraviolet Visible spectroscopy. The ionic conductivity of gel polymer electrolyte was studied by electrochemical impedance spectroscopy (EIS). NDSSCs was assembled according to the configuration of (i) FTO/TiO$_2$/natural dye/gel polymer electrolyte/carbon/FTO (ii) FTO/ TiO$_2$-CuO/natural dye/gel polymer electrolyte/carbon/FTO and (iii) FTO/TiO$_2$-ZnO/natural dye/gel polymer electrolyte/carbon/FTO were fabricated and results were compared. The solar cell efficiency has calculated using solar simulator AM 1.5 irradiation.

The metal oxide such as ZnO and CuO onto the TiO$_2$ nanoparticles, creating a core-shell nanoporous structure, provides means to minimise the charge recombination losses in DSSCs. The improvement in the solar cell energy conversion efficiency by the over coating approach may be assigned to the following factors: (i) the wide band gap coating delays the electron back transfer to the electrolyte and minimises charge recombination [11], (ii) the coating layer also enhances the dye adsorption onto the porous electrode and, as a consequence, the dye loading, increasing the photocurrent [12]. It was achieved by TiO$_2$-CuO and it shows better efficiency than the TiO$_2$-ZnO and TiO$_2$ photoanode.

2.2 Experimental

2.2.1 Materials

All chemicals used were purchased are of analytical grade and have used without purification. Copper nitrate (CuNO$_3$.3H$_2$O), zinc nitrate (Zn(NO3)$_2$.6H$_2$O), sodium hydroxide (NaOH), iodine (I$_2$) and potassium iodide (KI) were purchased from HiMedia. Acetonitrile was purchased from Merck. Titanium isopropoxide, PMMA, PEG (Mw 6000) and Triton X-100 were obtained from Aldrich. Acetonitrile, acetone, ethanol, Isopropanol and acetylacetone were purchased from Merck.

2.2.2 Methods

2.2.2.1 Preparation of nano-TiO$_2$, nano-ZnO and nano-CuO

The nano-TiO$_2$ (T1), nano-ZnO (Z1) and nano-CuO (C1) were prepared by sol-gel method using titanium isopropoxide, zinc nitrate hexahydrate and copper nitrate trihydrate as precursors, respectively [13, 14, 15].

2.2.2.2 Preparation of TiO$_2$/ZnO and TiO$_2$/CuO Core/Shell nanomaterials

Zinc nitrate hexahydrate (Zn(NO$_3$)$_2$.6H$_2$O), sodium hydroxide (NaOH) and presynthesised T1 nanoparticles were used for the preparation of TiO$_2$/ZnO core/shell nanomaterials. To 10 mM of (Zn(NO$_3$)$_2$·6H$_2$O), was dissolved in deionised water with constant stirring for 10 min; then 10% of presynthesised TiO$_2$ nanoparticles were added to it. To the above solution 0.1M NaOH solution was added dropwise and the resulting solution was stirred for another 3 hrs. Finally the resulted solution was refluxed at 97°C for 3 h to form ZnO shell on the surface of TiO$_2$ nanoparticles. Then it was cooled down to room temperature, centrifuged, washed with deionised water and acetone. It was dried at room temperature to obtain TiO$_2$/ZnO core/shell nanomaterials. Here, Ti^{2+}/Zn^{2+} were taken as 90:10 and identified as T1Z1. Similarly, we also prepared in the ratio of 80:20, identified as T1Z2. The final products of all samples were annealed in air at 450°C for 2h for further characterisation and used as photoanode material in the fabrication of NDSSC.

The above procedure was repeated using copper nitrate trihydrate to get TiO$_2$/CuO core shell nanomaterials. They were identified as T1C1 (90:10) and T1C2 (80:20). The final products of all samples were annealed in air at 300°C for 4 h.

2.2.3 Fabrication of DSSC Electrodes

2.2.3.1 Preparation of photoanode

Fluorine Tin Oxide (FTO) glass was cleaned using acetone, deionised water and isopropanol for regular interval time of 15 min under sonication. The cleaned FTO glass plate was preserved in isopropanol solution before use. In this study, we assembled three DSSC which consist of three different photoanode such as TiO$_2$ (T1) and prepared Core/Shell nanomaterial TiO$_2$/ZnO (**T1Z2**) and TiO$_2$/CuO (**T1C2**) using doctor blade method [16]. An active area of 0.5x0.5 cm^2 was identified on the FTO-glass substrate (sheet resistance 13 Ω/cm^2) and the T1 paste was dropped on the conductive side of FTO glass plate using glass rod. The film was dried in an air at room temperature. After 20 min it was sintered about 30 min at 450°C. Then, it was cooled in an oven till it reaches 80°C. After the attainment of 80°C, the film was immersed in ethanolic solution of natural dye for 24 h at room temperature to dye sensitised photoelectrode.

2.2.3.2 Preparation of gel polymer electrolyte

An optimised composition of (60:40 wt%) of PMMA and PEG were dissolved under continuous stirring in acetonitrile at 60°C. After about 2 h, 10 wt% KI and 1 wt% I_2 were added to the above polymer solution to form polymer blend electrolyte (PE). The polymer solution was stirred continuously for overnight to get homogeneous viscous gel. The solvent free polymer nanocomposite standing film was obtained by drying the resulting solution in an oven at 60°C for 24 h. The dried films were packed and stored in dark desiccators.

2.2.3.3 Preparation of natural dye sensitiser

Fresh Cactus pear fruit were obtained from Uthukottai taluk, Thiruvallur district, Tamil Nadu. The fruits were washed with distilled water and the outer skin and the seeds were removed. Then the betalain dye extracted from fruit pulp was stirred with 100 ml of ethanol at room temperature. The pH was adjusted to 2 by adding 10 mM of HCl. Then the solid dregs are filtered and obtain a pure natural dye solution. The residual parts removed by filtration were washed with hexane several times to remove any oil or chlorophyll present in the extract. This is directly used as a dye solution for sensitising three photoanodes (T1, T1Z2, T1C2) electrodes.

2.2.3.4 Preparation of counter electrode

Counter electrode was prepared by depositing carbon soot on the conductive side of glass with the help of candle flame. Substrate was passed over the tip of candle flame for 2 minutes with repeated back and forth movement.

2.2.3.5 Cell assembly

For two-electrode measurement, the electrolyte was sandwiched between photoanode and platinum-coated counter electrode, pressing firmly. A thin layer of paraffin was used as a spacer to avoid short-circuiting between two electrodes. A binder clip was fixed externally to maintain the mechanical grip of the cell without any further sealing, which finalised the assembly of the DSSC. Three NDSSCs was assembled according to the configuration of FTO/TiO$_2$/natural dye/gel polymer electrolyte/carbon/ FTO, FTO/T1C2/natural dye/gel polymer electrolyte/carbon/FTO and FTO/T1Z2/natural dye/gel polymer electrolyte/carbon/FTO.

2.2.4 Characterisation Methods

A Siemens D 500 X-ray diffractometer with CuKα filtered radiations was used to determine the structure and crystallinity of the samples. The shapes and surface morphology were studied using an S-4800 high-resolution scanning electron microscope (HRSEM). The composition of various core shell nanomaterial were scrutinised by energy dispersive X-ray spectroscopy (EDS). The conductivity of polymer electrolyte compositions measured using the HIOKI 3532 LCR Hi-Tester at room temperature. The conductivity was calculated using equation: $\sigma = t/A \times R_b$. Here R_b is the bulk resistance (Ω or S^{-1}) of the sample taken at the intersection of the Nyquist plot with the real impedance axis, t is the sample thickness (cm) and A is the surface area of the electrode/electrolyte contact (cm^2). The second part of the study was focussed on the fabrication and characterisation of the DSSCs. The absorption spectra of betacyanin dye were taken with a Shimadzu UV–1650PC UV-Vis spectrophotometer.

Photocurrent - voltage characteristics was measured using Oriel Class-A Simulator (M-91900 A, Newport) with Xenon lamp as a light source, having an intensity of 100 mW/cm^2. A computer controlled Auto lab PGSTAT302N electrochemical workstation was used for current- voltage measurements.

The fill factor (FF) is calculated from the I-V curve using the following equation:

$$FF = (I_{max} \times V_{max})/(I_{sc} \times V_{oc}) \tag{2.1}$$

where I_{max} and V_{max} denote the maximum output value of current and voltage, respectively, and I_{sc} and V_{oc} denote the short circuit current and open-circuit voltage, respectively. The overall energy conversion efficiency (η) is defined as,

$$\eta = (I_{sc} \times V_{oc} \times FF)/P_{in} \tag{2.2}$$

where, P_{in} is the power of incident light.

2.3 Results and Discussion

2.3.1 UV-Visible Spectroscopy

The UV-visible absorption spectrum of cactus fruit (*Opuntia stricta*) dye extract from ethanol is shown in the Figure 2.1. The broad absorption and sharp intense peak observed at 535 nm corresponds to betalain pigment.

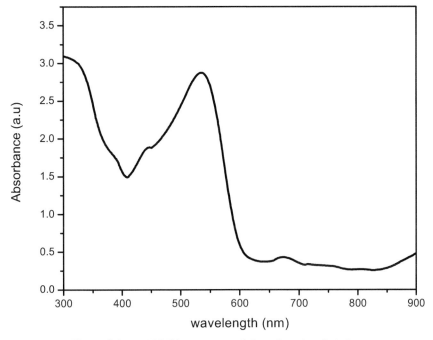

Figure 2.1   UV-Visible spectrum of *Opuntia stricta* fruit dye.

2.3.2 Fourier-Transform Infrared (FTIR) Spectroscopy

FTIR spectra of sample T1, Z1, C1 are given in Figure 2.2. It shows a broad band at 400-700cm^{-1}. This is attributed to Ti-O stretching and Ti-O-Ti bridging stretching mode [17]. The samples show peaks corresponding to stretching vibration of O-H and bending vibrations of adsorbed water molecules around 3300-3450 cm^{-1} and at 1640 cm^{-1}, respectively [18]. The characteristic absorption peak at 476 cm^{-1} due to Zn-O vibration authenticates the presence of ZnO. The FTIR spectrum of n-CuO shows bands at around 615 and 506 cm^{-1}, which can be assigned to the vibrations of Cu(II)-O bonds. There is sharp peak observed at 506 cm^{-1} in the spectrum CuO nanoparticles which is the characteristics of Cu-O bond formation. In T1Z1 and T1Z2, a sharp strong band was observed around 471 cm^{-1} which corresponds to stretching mode of Zn-O nanoparticles. With the increase of Zn content Ti-O bands shift to lower wavenumber region and gets sharpened [19]. Whereas for T1C1 and T1C2, Ti-O bands shift to higher wavenumber region and gets sharpened.

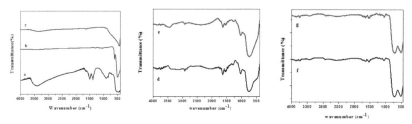

Figure 2.2   FTIR spectra of (a) Z1, (b) C1, (c) T1, (d) T1Z1, (e) T1Z2, (f) T1C1 and (g) T1C2.

2.3.3 Scanning Electron Microscopy (SEM)

Figure 2.3 shows the HRSEM micrographs of the synthesised samples, T1 shows approximately spherical morphology with the agglomeration of particles of 10–15 nm, which is the core material. Pure ZnO (Z1) shows nanorods with 15-45 nm. Pure CuO (C1) shows 9–16 nm. With the increase of ZnO and CuO coating on the core material, the size of the particle gets slightly increased as 15–19 nm for T1Z1, 24–36 nm for T1Z2, 50–70 nm for T1C1 and 35–80 nm for T1C2 which is related to the shell thickness, respectively, as presented in Figure 2.3 (d-g). The HRSEM image of polymer gel electrolyte is shown in Figure 2.3 (h) the smooth surface of GPE.

2.3.4 Energy Dispersive Spectroscopy (EDS)

The chemical compositional analysis of the samples is very essential to know the exact concentration of elements, added dopants, and defects if any present in samples. The EDS spectra of Z1 (Zn, O), C1 (Cu, O) and T1 (Ti, O) shown in Figure 2.4 (a, b and c) indicates the purity of the prepared samples. ZnO-coated TiO_2 (Figure 2.4d, e) and CuO-coated TiO_2 (Figure 2.4 f, g) systems, Zn and Cu are also present along with Ti and O elements, respectively, which confirms the absence of any other impurities. The elemental composition of Z1, C1, T1, T1Z1, T1Z2, T1C1 and T1C2 are given in Table 2.1.

2.3.5 Electrochemical Impedance of Gel Polymer Electrolyte

The ionic conductivity of gel polymer electrolyte (PMMA/PEG/KI/I_2) is found to be 3.25×10^{-6} S cm^{-1}.

2.3.6 Current Voltage Characteristics

The current density-voltage (J-V) characteristics of DSSCs prepared using T1, T1Z2 and T1C2 nanoparticles as photoanode are shown in the Figure 2.6.

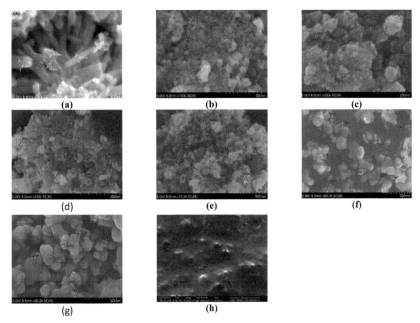

Figure 2.3 HRSEM images of (a) Z1, (b) C1, (c) T1, (d) T1Z1, (e) T1Z2, (f) T1C1, (g) T1C2 and (h) GPE.

Table 2.1 elemental compositions of prepared various core shell nanomaterial.

Sample	Element	Weight%	Atomic %	Sample	Element	Weight%	Atomic %
Z1	Zn L	72.16	38.81	T1Z2	O K	46.49	73.19
	O K	27.84	61.19		Ti K	38.66	20.86
C1	O K	35.24	68.37		Zn L	14.85	5.95
	Cu L	64.76	31.63	T1C1	O K	42.10	69.32
T1	O K	41.09	67.62		Ti K	49.30	27.11
	Ti K	58.91	32.38		Cu L	8.60	3.57
T1Z1	O K	44.56	71.60	T1C2	O K	44.50	72.09
	Ti K	45.97	24.68		Ti K	39.61	21.43
	Zn L	9.47	3.72		Cu L	15.88	6.48

The measurements are carried out by using two electrode systems. Xenon lamp was used as a light source and the incident light was maintained at $100W/cm^2$. The significant changes associated with the T1Z2 and T1C2 are mainly due to the open circuit voltage (V_{oc}) values. The core/shell electrodes provide an inherent energy barrier at the electrode-electrolyte interface, thereby the electron recombination loss are reduced. Hence, V_{oc}

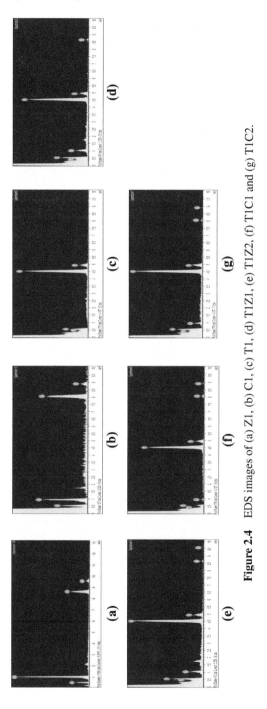

Figure 2.4   EDS images of (a) Z1, (b) C1, (c) T1, (d) T1Z1, (e) T1Z2, (f) T1C1 and (g) T1C2.

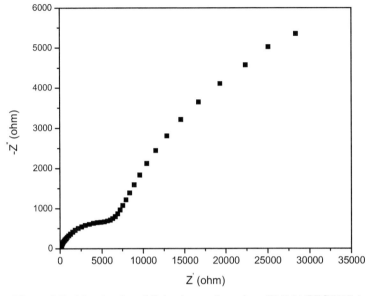

Figure 2.5 Nyquist plot of Gel polymer electrolyte (PMMA/PEG/KI/I$_2$).

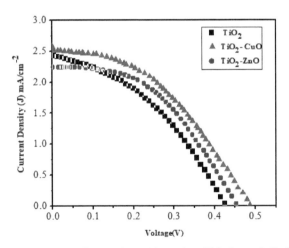

Figure 2.6 J-V curves of DSSCs based on TiO$_2$ and on TiO$_2$/core shell electrodes and gel polymer electrolyte under 1 sun (100 mWcm^{-2} AM 1.5).

value increased. The higher V$_{oc}$ values observed in T1C2 based core/shell electrode arise from the conduction band edge shift of the TiO$_2$ brought by coating the electrode with a different oxide endowed with higher conduction band energy.

Table 2.2   J-V features from the DSSCs based on TiO_2 and TiO_2 core/shell electrodes and gel polymer electrolyte under 1 sun illumination (100 mWcm^{-2} AM 1.5).

Electrode	V_{oc} (V)	J_{sc} (mA/cm^2)	V_{max} (V)	J_{max} (mA/cm^2)	FF	η (%)	Reference
TiO_2	0.43	2.43	0.24	1.69	0.39	0.41	Present work
TiO_2CuO	0.49	2.53	0.27	1.90	0.41	0.51	Present work
TiO_2 ZnO	0.45	2.24	0.26	1.78	0.46	0.46	Present work
TiO_2 ZnO	0.41	2.77	0.31	1.32	0.354	0.41	[20]

In comparison with the results, J_{sc} values are too small for T1Z2 due to small particle size which does not support for space charge region and it led to the recombination of photogenerated charge carriers. Another reason is that it provides interfacial recombination in individual core material.

2.4 Conclusion

Two core/shell electrode was prepared by using refluxing process for the formation of ZnO and CuO on the surface of chemically pre-synthesised TiO_2 nanoparticles. The SEM with EDS analysis showed spherical morphology with agglomeration of nanoparticles and the presence of Ti, Zn, Cu and O elements alone in samples. From the results of J-V characterisation, it was inferred that the addition of ZnO layer increased V_{oc} from 0.43 V to 0.45 V for ZnO-coated TiO_2 and 0.43V to 0.49 V, confirming its active role as an inherent energy barrier that leads to the decrease in recombination losses. On comparing the efficiency of three NDSSCs, the core/shell material of TiO_2/CuO shows increased efficiency from 0.41% to 0.51%.

Acknowledgements

We are gratefully acknowledging the financial support of DST INSPIRE (IF 140656) (No. DST/INSPIRE Fellowship/2014/269).

References

[1] J. Liu, Y. Zhao, A. Wei, Z. Liu, F. Luo, J. Mater. Sci. Mater. Electron. (2014), 25, 4008.
[2] B. O'Regan, M. Gratzel, Nature (1991), 353, 737.

[3] A. Yella, H.-W. Lee, H.N. Tsao, C. Yi, A.K. Chandiran, M.K. Nazeeruddin, E.W.-G. Diau, C.-Y. Yeh, S.M. Zakeeruddin, M. Grätzel, Science, (2011), 334, 629.

[4] V. Thavasi, V. Renugopalakrishnan, R. Jose, S. Ramakrishna, Mater. Sci. Eng.: R:Rep. (2009), 63, 81.

[5] Y. Xu, T. Gong, J.N. Munday, Sci.Rep. (2015), 5, 13536.

[6] V. Manthina, J.P. Correa-Baena, G. Liu, A.G. Agrios, J. Phys. Chem. C. (2012), 116, 23864.

[7] S.S. Kanmani, K. Ramachandran, Renew. Energy, (2012) 43, 149.

[8] Y. Wang, Y.-Z. Zheng, S. Lu, X. Tao, Y. Che, J.-F. Chen, ACS Appl. Mater. Interfaces (2015), 7, 6093.

[9] D. Guo, J. Wang, C. Cui, P. Li, X. Zhong, F. Wang, S. Yuan, K. Zhang, Y. Zhou, Sol. Energy (2013), 95, 237.

[10] R. Elangovan, N.G. Joby, P. Venkatachalam, J. Solid State Electrochem. (2014), 18, 1601.

[11] E. Palomares, J.N. Clifford, S.A. Haque, T. Lutz, J.R. Durrant, J. Am. Chem. Soc. (2003), 125, 473.

[12] S. Lee, J.Y. Kim, K.S. Hong, H.S. Jung, H. Shin, Sol. Energy Mater. Sol. Cells (2006), 90, 2405.

[13] Choi HC, Ahn HJ, Jung YM, Lee MK, Shin HJ, Kim SB, et al., Appl Spectrosc (2004), 58, 598-602.

[14] Hari Hara Priya G, Suganya N, Jaisankar V, Int. J. Chemtech Res., (2015), 7, 2942-2948.

[15] Prakash Chand, Anurag Gaur, Ashavani Kumar, AIP Conf. Proc., 1393, (2011), 211-212.

[16] R. Adel, T. Abdallah, Y.M. Moustafa, A.M. Al-sabagh and H. Talaat, Super lattices Microstruct., (2015), 86, 62–67.

[17] Yu JG, Yu HG, Cheng B, Zhao XJ, Yu JC, Ho WK. J. Phys. Chem. B (2003), 107, 1387.

[18] Hamadanian M, Reisi-Vanani A, Majedi A, J. Iran Chem. Soc., (2010), 7, 52.

[19] Wang Z, Zhang H, Zhang L, Yuan J, Yan S, Wang C, (2003), 14,115.

[20] S. Sakthivel and V. Baskaran, Nano Vision, (2015), 5, 169–294.

3

Implementing ZnO Nanomaterials in P3HT:PCBM Based Hybrid Solar Cell

R. Geethu[1,*,†], K. S. Ranjith[2,‡], R. T. Rajendra Kumar[3], and K. P. Vijayakumar[1,§]

[1]Thin Film Photovoltaic Division, Department of Physics, Cochin University of Science and Technology, Cochin, Kerala, India
[2]Advanced Materials and Devices Laboratory, Department of Physics, Bharathiar University, Coimbatore, India
[3]Department of Nanoscience and Technology, Bharathiar University, Coimbatore, India
E-mail: geethumangalath@gmail.com
*Corresponding Author

Objectives

- Develop nanomaterials like nanorods, nanoroots through simple and economic routes.
- Characterise these synthesised nanomaterials
- Utilising these prepared ZnO nanomaterials for solar cell fabrication and analyse device performances
- Compare the photovoltaic parameters of different devices

This chapter focuses on fabricating hybrid type solar cell using nanomaterials of ZnO as the electron collective layer. Nanomaterials, vertically aligned ZnO nanorods and ZnO nanoroots were developed through simple, cost-effective and non-vacuum aqueous solution growth and chemical

---

†Present address: Department of Applied Chemistry, Cochin Universityof Science and Technology, Cochin-22, Kerala
‡Department of Energy and Materials Engineering, Dongguk University, Seoul 04620, Republic of Korea
§Center of Excellence in Advance Materials, Cochin University of Science and Technology, Kochi, Kerala

spray pyrolysis techniques. Seed layer is required for the growth of vertically well-aligned ZnO nanorods and was deposited using chemical spray pyrolysis technique. Structural, optical and morphological characterisations revealed the formation of ZnO film with uniform surface when deposited at substrate temperature of 550 °C. This also described the role of pH, Zn precursor concentration for the growth of vertically aligned ZnO nanorods. Another type of nanostructure, that is, tangled root structure, was developed through chemical spray pyrolysis and was morphologically analysed. Solar cells were fabricated with basic device structures: ITO/ZnO:Al (seed layer)/ZnO (nanorods)/P3HT:PCBM/Ag, ITO/ZnO (tangled roots)/P3HT:PCBM/Ag. Modifications in these structures were also done. Devices fabricated with modified initial structure exhibited an efficiency of ∼0.44 % and that with the later yielded an efficiency of ∼1.4 %.

3.1 Introduction

Research on nanomaterials is increasing day by day because of the application in various fields like healthcare (biosensing, tumour detection, bioimaging, nanozymes), filters (nanostructured filters for water and air purification), paints (enhance protection from UV) and lubricant additive (reduce friction of moving parts) [1-4]. They also have applications in the field of construction (like nanocoating for concrete, improve durability of wood), nanocomposites, etc. [5-6].

Use of nanomaterials in the field of solar energy conversion is also inevitable. Even though potential of solar energy to solve the energy crisis is well-known, trapping of sunlight reaching the earth atmosphere is not efficient yet. In particular case of solar cell too, light trapping is challengeable. Here comes one of the applications of nanomaterials in solar cells, that is, light harvesting. Moreover, these nanosized particles can provide higher surface area, creating more active sites and also they obstruct the recombination of photo generated charge carriers [7].

Nanomaterials are mainly defined in terms of size, that is, those are materials possessing size in the range 1-100 nm or have nanoscale dimension [8]. Utilisation of nanomaterials for optoelectronic application now became common because of their interesting optical and electrical properties. Tunability of their optoelectronic properties is another reason for the interest [9-10]. Incorporation of nanomaterials in a photovoltaic device enhances the device performance [11-13]. Commonly used nanomaterials include nanorods, nanowires, nanoparticles, nanoflowers, quantum dot, nanostructured arrays etc. [14-16].

Commonly used nanomaterials as electron transport layer in organic solar cells are ZnO or TiO$_2$ [17-18]. Comparing to TiO$_2$, ZnO possesses several advantages like high electron mobility, easily deposited through different techniques, low reflectivity, high surface to volume ratio etc. which makes them good candidate for solar cell application [19-21]. Research works were already reported based on inserting nanostructured layers or mixing active layers with nanoparticles [22-38]. Hu et al., fabricated the solar cell with device structure FTO/ZnO/P3HT:PCBM/Ag. With this basic configuration, three independent type of device were fabricated with different electron collective layers as, (a) ZnO polycrystalline seed layer, (b) loose nanopillars atop polycrystalline seed layer and (c) dense nanopillars atop polycrystalline seed layer. Better performance was exhibited by the device which utilise loose nanopillars and yielded an efficiency of ~1 %, which might be due to larger interfacial area between ZnO and polymer provided by them [26]. Tanveer et al. developed photovoltaic devices with comparable configuration, but with nanofibres instead of nanopillars. Without nanofibers device attained an efficiency of 0.9 % whereas utilising nanofibres solar cell efficiency reaches up to 2.23 % [27]. Another report by Choi et al., explains device fabrication using nanoparticles, nanorods and combination of both. Devices had MoO$_3$ (hole collective layer) over active layer. Those studies were also carried out in Argon filled gloves box [28]. Tong et al., reported a flexible polymer solar cell with configuration PET/ITO/ZnO thin film/ZnO nanorods/P3HT:PCBM/Ag in which they attained efficiency of 1.78 % [29]. Solar cells were fabricated with configuration ITO/ZnO/PFN/PTB7:PC71BM:ICBA/MoO$_3$/Ag by C. Han et al., to enhance the device performance. Combination of ZnO nanoparticles and PFN act not only as efficient electron transport layer for reducing series resistance and improving shunt resistance but also as effective surface modifier for better energy level alignment. These devices had efficiency of 9.31 % [37].

Research work presented here is focused on developing ZnO nanomaterial and applying it for solar cell application. ZnO nanomaterials were developed in two different ways and were incorporated with organic materials to form hybrid type device.

3.2 Vertically Well-aligned ZnO Nanorods and Its Solar Cell Application

Our aim was to grow vertically aligned ZnO nanorods. For the growth of nanorods a base layer called "seed layer" is required. Here uniform thin film

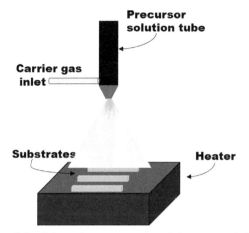

Figure 3.1   Schematic representation of the spray technique.

with lower resistivity was preferred to act as seed layer. Detailed description about the deposition of seed layer is discussed below.

3.2.1 Seed Layer Deposition

ZnO itself was chosen as the material to act as seed layer for the growth of ZnO nanorods. Since ZnO is a well-studied material, a wide variety of techniques are available for its deposition. Among chemical and physical methods chemical route was chosen since it is more cost effective than the other. A number of chemical methods like chemical bath deposition (CBD), sol-gel deposition, spin coating, chemical vapour deposition, chemical spray deposition etc. are available. From these, technique chosen to deposit ZnO seed layer was chemical spray pyrolysis (CSP). CSP is a simple, economic, non-vacuum technique through which uniform thin films can easily be deposited [39]. Schematic diagram of spray technique is shown in Figure 3.1. As for all techniques, spray pyrolysis has some basic parameters that determine nature of the formed thin film. These are carrier gas pressure, substrate temperature, spray rate, time period of spray and molarity of the solution. In CSP technique precursor solution is loaded into the container and substrates were kept on a heater. Predetermined quantity of solution reaches the nozzle through a tube from the container. Inside the nozzle solution get mixed with the carrier gas and is sprayed out of the nozzle towards the substrate.

For the deposition of ZnO seed layer, in the present work precursor solution was prepared using zinc acetate, water and propanol. Molarity of the

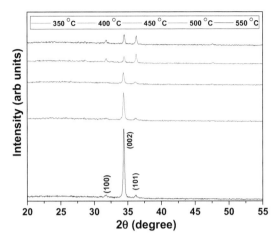

Figure 3.2 XRD profile of the samples deposited at different substrate temperature [40].

precursor solution was 0.3 M. Equal ratio of water and propanol were taken for the dissolution of precursor salt and hydroxide formation was nullified by adding a few drops of acetic acid. Soda lime glass - microscopic slide: $1'' \times 3''$; 1.2 mm thickness were used as the substrates. For the deposition of thin film using CSP technique, substrates were kept on a heater. Temperature of the substrate plays a vital role in the formation of thin film with uniform surface. Moreover, it has major role in determining the growth of vertically aligned ZnO nanorods. Hence, this is discussed here in detail by keeping other parameters spray rate, carrier gas pressure, volume of the precursor solution as 7 ml/min, 20 psi and 70 ml, respectively. Substrate temperature was varied from 350 to 550 °C in the steps of 50 °C. Variation in ZnO thin film properties with substrate temperature can be analysed from the characterisation studies done for those samples.

Structural characterisation was done by taking X-ray diffraction and was done using Rigaku D Max C X-ray diffractometer with Cu K_α line as the source radiation along with Ni filter. This was operated at voltage of 30 kV and current of 20 mA. XRD pattern of the sample is in Figure 3.2. ZnO thin films deposited at all temperatures has three peaks corresponding to the diffraction from (100), (002) and (101) plane. Comparing XRD pattern of the samples deposited at different temperature intensity of peak corresponding to the reflection from (002) increases with temperature [40]. Structural characterizisation reveals the formation hexagonal wurzite structure of ZnO thin films.

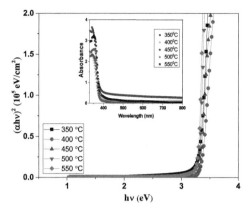

Figure 3.3 Band gap determination for the samples deposited at different substrate temperature. Inset shows the optical absorption spectra of the samples.

Optical studies were done by using UV-Vis-NIR spectrophotometer (JASCO-570 model). Absorption coefficient can be determined using the equation

$$T = (1 - R)^2 \exp(-A) = (1 - R)^2 \exp(-\alpha t)$$

where T is the transmittance of the films, R is the reflectance, A is the absorbance, α is the absorption coefficient and t is the film thickness [41]. Band gap (E_g) was determined from the equation

$$\alpha h\nu = A (h\nu - E_g)^\gamma$$

where γ is a constant which is equal to 1/2 and 3/2 for allowed and forbidden direct transitions, respectively, and equal to 2 and 3 for allowed and forbidden indirect transitions in which the phonons are involved, A is a constant and E_g is the optical band gap [41]. Tauc plot ($h\nu$ vs. $(\alpha h\nu)^2$) is used to determine the band gap of ZnO thin film since ZnO is a direct band gap material. Figure 3.3 shows optical band gap and its inset shows the absorbance spectra of ZnO sample deposited at different substrate temperature. Optical absorption spectra gives the band gap of the deposited samples and were found to ~3.3 eV.

Morphological characterisation was done by taking AFM measurements (Figure 3.4). It shows microscopic view of sample surface. Comparing RMS value of surface roughness of the samples at different substrate temperature, it was higher at lower temperature (~170 nm) and decreases to < 30 nm for samples prepared at 450, 500 and 550 °C.

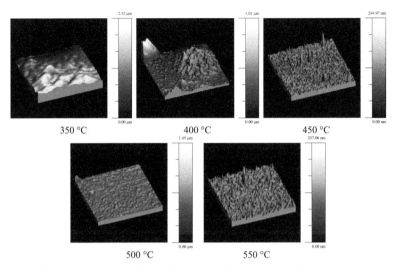

Figure 3.4   AFM images of samples deposited at different substrate temperature.

To study how the seed layer substrate temperature affects the growth of nanorods, ZnO nanorods were grown on samples deposited at different substrate temperature and are discussed below.

3.2.2 Growth of Vertically Well-aligned ZnO Nanorod

ZnO nanostructures were synthesised by a low temperature aqueous solution growth method [42]. Growth solution was prepared by dissolving zinc nitrate hexahydrate $Zn(NO_3)_2.6H_2O$ and hexamine (HMTA) ($C_6H_{12}N_4$) in double distilled water. This was stirred continuously for 15 min in separate beakers. For the formation of single phase solution HMTA solution was added drop wise to $Zn(NO_3)_2$ solution and stirred continuously for 20 min. The seed layer coated substrates were immersed in the growth solution for 4 hours at a temperature of 90 °C. After completing the growth process, the substrates with deposited nanostructures were cleaned twice, with distilled water and ethanol, and baked at 150 °C.

3.2.2.1 Role of seed layer deposition temperature on the growth of vertically aligned ZnO nanorods

As discussed above nanorods were grown over seed layer deposited at different substrate temperature (350, 400, 450, 450, 500 and 550 °C). At lower substrate temperatures nanowires were formed instead of nanorods and the growth of nanorods became more prominent at higher temperatures. On

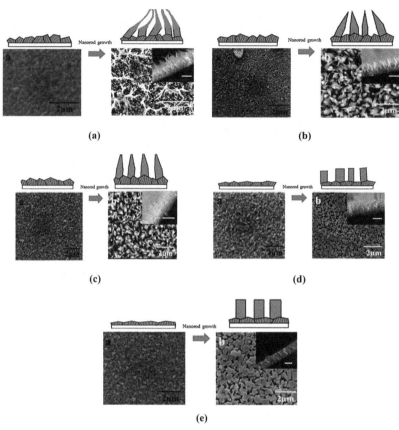

Figure 3.5   Topological SEM images of ZnO seed layer and ZnO nanorods at different seed layer deposition temperature (a) 350 °C, (b) 400 °C, (c) 450 °C, (d) 500 °C, (e) 550 °C. Picturisation of sample surface is drawn on the top of each image [40].

analysing the growth of nanorods from lower temperature, it was revealed that interconnected nanowires (formed at 350 °C) transforms to nanorods with reducing diameter towards the top (at seed layer substrate temperature 400 °C) reform as a nanorods with uniform diameter at higher seed layer substrate temperature (500 and 550 °C). Comparing the nanorods grown at these two higher temperatures, vertically well-aligned ZnO nanorod occurred at 550 °C. Detailed description about the growth of ZnO nanorods at different substrate temperature of seed layer and the precise reason for the vertical growth at higher temperature were discussed elsewhere [40]. SEM images of ZnO seed layer samples deposited at different substrate temperature and the nanorods grown over it are given in Figure 3.5.

3.2.2.2 Role of pH, Zn precursor concentration for the growth of vertically aligned ZnO nanorod arrays

For optimising the morphology and alignment of the nanorod arrays, the reaction parameters like pH of the solution (pH 3 to 9 in the steps of 2) and ionic concentration (0.125 mM, 0.100 mM and 0.075 mM) [43] were varied. The growth mode was optimised as equal mole (0.125 mM) of $Zn(NO_3)_2.6H_2O$ and HMTA ($C_6H_{12}N_4$) as the growth precursors dissolved in double distilled water and stirred continuously for 15 min in separate beakers. HMTA solution was added drop wise to $Zn(NO_3)_2$ solution and the final solution was stirred continuously for 20 min to form single phase solution. To control the release of Zn ions in growth solution, initial growth solution was maintained at 5.5 pH using 50 μL of nitric acid. After immersing the nucleated substrate facing down using substrate holder, growth solution was maintained at 97 °C for 4 hours. After the growth time, ZnO nanorod array grown glass substrates were taken and washed with DI water to remove the unwanted residual zinc salt impurities and dried in a stream of N_2 at 150 °C. For improving the aspect ratio of the nanorod arrays, the growth was repeated 2-4 times and for improving the crystallinity without change in defect states, nanorod arrays were annealed at 400 °C under vacuum. Another factor that determines the growth of nanorods was molar ratio between the hexamine and zinc nitrate hexahydrate. Variation in this was already reported elsewhere [42].

pH of the solution was varied from 9 to 3 for the growth of nanorods on the optimised nucleated seed layer. Figure 3.6a–d shows the morphology of the ZnO nanorods on varying the pH of growth solution. Solution pH of 9 yielded nanoflowers like structure with size ~20 nm whereas morphological image of sample at pH 7 reveals the formation of nanorods with diameter in the range 50-100 nm. Reducing pH to 5 gave uniform and aligned ZnO nanorods possessing a diameter of 80 nm. Further reduction of pH to 3 yielded mixture of nanorods (with diameter 20-30 nm) and 2D nanosheets (shown in Figure 3.6d) [43]. pH 5 is suitable condition for the growth of vertically aligned nanorod arrays with uniform density which are parallel to the substrate [44].

Morphological variation in nanorod arrays for zinc precursor concentrations 0.125 M, 0.100 M and 0.075 M are shown in Figure 3.7a–c. Precursor concentration of 0.125 M yielded well faceted hexagonal structure where ZnO nanorods lost it for higher precursor concentration of 0.100 M. ZnO nanorods with sharp tips were obtained for precursor concentration of

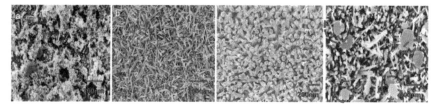

Figure 3.6   SEM image of ZnO nanostructured arrays grown on seed substrate with different pH values: (a) pH 9, (b) pH 7, (c) pH 5 and (d) pH 3 [43].

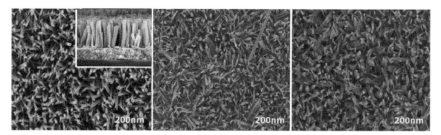

Figure 3.7   SEM image of ZnO nanostructured arrays for zinc precursor concentrations (a) 0.125 M, (b) 0.100 M and (c) 0.075 M [43].

0.075 M. Comparing these three, better nanorods array with uniform diameter throughout its length was formed at 0.125 M precursor concentration. Detailed description is given elsewhere [43].

3.2.3 Hybrid Solar Cell Fabrication Using ZnO Nanorods

Solar cell fabrication includes different stages of deposition process. In this case, device fabrication includes four stages of deposition process. ZnO seed layer deposition over ITO (Tin doped indium oxide, Geomatec Japan), ZnO nanorods deposition over seed layer, organic layer deposition over ZnO nanorods and electrode (silver metal) deposition over organic layer. First two steps are already discussed above. Third step and fourth steps are discussed below.

3.2.3.1 Organic layer deposition

Organic layer chosen was a blend of a polymer and an organic molecule. Polymer chosen was P3HT [poly(3-hexylthiophene) - regioregular] (purchased from Rieke metals) and organic molecule is PCBM ([6,6]-phenyl-C_{60}-butyric acid methyl ester) (purchased from Ossila). Structures of both are as

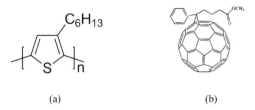

(a)   (b)

Figure 3.8   Structure of (a) P3HT and (b) PCBM.

shown in Figure 3.8. These two materials, the polymer which act as donor and the molecule which act as acceptor are blended together and are deposited as single film. This single thin film in an organic solar cell is called as active layer since it consist of both donor and acceptor materials. Blending polymer and molecule can be done by dissolving powders of two materials in required ratio using common organic solvent (like chloroform, chlorobenzene and dichlorobenzene). For this particular study, P3HT and PCBM were taken in 1:1 ratio and were dissolved in chlorobenzene. Sonication was also provided for complete dilution.

Technique chosen for the deposition of thin film was spin coating. It is one of the simplest techniques that utilise centripetal force for the deposition of uniform thin films [45]. Basic parameters of spin coating techniques that determines the quality of film formation are spinning speed, spinning time and spinning acceleration. Spin coating can be done in two different ways (1) by dropping the solution on to the spinning substrate (called as dynamic dispense spin coating technique) and (2) by dropping the solution on to the static substrate and then spinning (called as static dispense spin coating technique) [46]. Active layer deposition for this work utilised static dispense spin coating, in which solution is dropped on to the centere of the substrate kept on the chuck of the spin coater. Spinning the substrate causes the solution to spread over the substrate due to centripetal acceleration [47- 48] resulting in the formation of uniform thin film. A schematic diagram of spin coater is shown in Figure 3.9.

3.2.3.2 Top electrode deposition

Over active layer metal electrode was deposited by vacuum evaporation technique. Silver (Ag) was chosen as the top electrode and was deposited at a pressure of 10^{-6} torr. Thickness of silver electrode was $\sim$50 nm. Round electrodes with electrode area 0.03 cm^2 were deposited with the help of a mask.

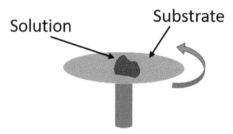

Figure 3.9   Schematic representation of spin coating technique [47].

3.2.3.3 Device characterisation

Solar cell performance can be analysed by measuring its voltage-current characteristics. Dark and illuminated I-V measurements can be done with the help of a source measure unit (it can measure current for different applied voltages or vice versa) along with a solar simulator (provide the light with intensity that matches the solar spectrum) for illumination. For this particular study source measure unit used was NI-PXI 1033, National instruments and the solar simulator was Class AAA (PET, USA, model-SS50AAA). When measured under dark, a solar cell shows a diode characteristics from which diode quality factor can be determined. Diode quality factor for the device was calculated from the diode equation [49].

$$I = I_0 \left[\exp\left(\frac{qV}{nkT} \right) - 1 \right]$$

I – current, I_0 – saturation current, q – charge of electron, V – voltage, n – diode quality factor, k – Boltzmann constant, T – temperature in Kelvin. In an ideal case, diode quality factor ~ 1.

On illuminating a solar cell, it displays an I-V characteristic similar to the plot shown in Figure 3.10. For a solar cell, equation can be formulated as [50].

$$I = I_0 \left[\exp\left(\frac{qV}{nkT} \right) - 1 \right] - I_L$$

where I – net current, I_0– saturation current, q- charge of electron, V – voltage, k – Boltzmann's constant, T- temperature in Kelvin, I_L– photo current.

Basic parameters of a solar cell obtained from I-V measurements are

(a) Open circuit voltage
(b) Short circuit current

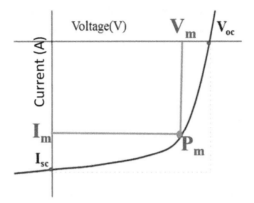

Figure 3.10 Illuminated I-V characteristics of a solar cell.

(c) Fill factor
(d) Efficiency

Open circuit voltage (V$_{oc}$): Voltage obtained from a solar cell at zero current or it is the voltage obtained when the circuit is opened. It is marked as V$_{oc}$ in Figure 3.10. It is the highest value of voltage that the solar cell can attain. Since at this point net current is zero, then V$_{oc}$ is [51]

$$V_{oc} = \frac{nkT}{q} ln \left(\frac{I_L}{I_0} + 1 \right)$$

From the equation it is clear that V$_{oc}$ depends on photo current and saturation current. Since I$_o$ depends on recombination, V$_{oc}$ is affected by the recombination process in the solar cell.

Short circuit current (I$_{sc}$): It is the current obtained from a device when the circuit is shorted. At this point, voltage is zero. In the figure, it is marked as I$_{sc}$. Light generated and collected charge carriers contribute to I$_{sc}$ and hence, for an ideal solar cell, I$_{sc} \sim I_L$, under most moderate resistive loss. Therefore, it is the maximum current obtained for a device. I$_{sc}$ depends on the following factors:

1. Device area: this dependence can be eliminated by considering short circuit current density (J$_{sc}$ in mA/cm^2) instead of I$_{sc}$
2. Power of the incident light (number of photons incident on the sample)
3. Spectrum of the incident light
4. Optical properties (absorption and reflection) of the solar cell

5. Collection probability of the solar cell, which depends mainly on the surface passivation and the minority carrier lifetime [52,48]

Fill factor (FF): To calculate fill factor (FF) maximum power obtained from the device has to be determined. At maximum voltage point (V_{oc}) current is zero and at maximum current (I_{sc}) voltage is zero. Hence, the maximum power point will lie somewhere in between these two points and let it be P_m (marked in Figure 3.10). Fill factor is then defined as

$$\text{Fill factor (FF)} = P_m/(V_{oc} \times I_{sc})$$

Product of V_{oc} and I_{sc} give the ideal power of the device, which never happens. Hence, FF can be defined as ratio of maximum power obtained from the device to the ideal power of the device or it is the 'squareness' of the curve of solar cell [53, 48].

Efficiency (η): Efficiency can be defined as the ratio of output power (P_m) over input power (P_{in}). Input power is the product of power density of light incident on a solar cell and illuminated area of the device. Under standard condition (sun light falling on the earth's atmosphere during noon time, on a clear day), input power density is 100 mW/cm^2 or 1000 W/m^2. Efficiency can be formulated as

$$\eta = \frac{P_m}{P_{in}} \times 100$$
$$\eta = \frac{V_m \times I_m}{P_{in}} \times 100$$
$$\eta = \frac{V_{oc} \times I_{sc} \times FF}{P_{in}} \times 100$$

Initially solar cells were fabricated using device structure ITO/ZnO/P3HT:PCBM/Ag. ZnO chosen was same as that of ZnO seed layer. But the device did not exhibited a proper junction. When nanorods are incorporated the device structure became ITO/ZnO seed layer/ZnO nanorods/P3HT:PCBM/Ag (Figure 3.11).

This device shows diode characteristics without photo activity. Inefficient carrier collection might be a reason for this. Reducing the resistivity of seed layer ZnO might solve this issue. Reducing the resistivity of ZnO film can be done by in-situ doping of Al along with sample inversion [54]. Sample inversion means inverting the sample onto a surface immediately after film deposition to block direct contact of atmospheric oxygen. This process can reduce the ZnO resistivity by two orders [55]. Here ZnO resistivity after Al doping and inversion (ZnO:Al) reduces to 2×10^{-2} Ωcm from 8 Ωcm.

When devices were fabricated by replacing ZnO seed layer by ZnO:Al seed layer, they exhibit slight photo activity. This could be due to reduced

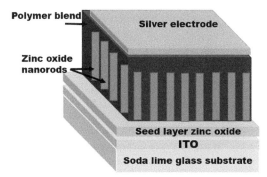

Figure 3.11 Schematic representation of the device using ZnO nanorods.

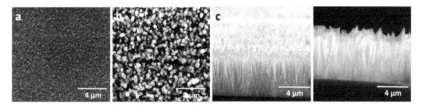

Figure 3.12 SEM images of (a) ZnO:Al, (b) and (c) top and cross-sectional view of ZnO nanorod arrays and (d) cross-sectional view of ZnO nanorods/P3HT:PCBM.

resistivity of the seed layer which initiated the charge carrier collection. J-V characteristic of the device is shown as inset in Figure 3.14. Open circuit voltage (V_{oc}) obtained was ~173 mV and short circuit current density (J_{sc}) was ~0.16 mA/cm^2.

Cross section and top view of SEM image of ZnO:Al (buffer layer) and ZnO nanorod arrays is shown in Figure 3.12a and b. The spray seed ZnO:Al thin film possessed uniform 2D assembly with controlled grain distriubution with the size of 200 nm. As utilising this buffer layer as nucleation platform, ZnO nanorod arrays were uniformly grown along the c axis with the length of 4 μm with the diameter of 200 nm. Figure 3.12d reveals the spin coating of P3HT:PCBM over the ZnO nanorod arrays provide the uniform top layer which has also fill the void of the nanorod arrays and provide the much interactive surface.

Performance of the device was low and to enhance it a hole collecting layer was inserted between active layer and silver electrode. Hole collecting layer chosen was PEDOT:PSS [poly(3,4-ethylenedioxythiophene):polystyrene sulphonate] and the device structure is shown in Figure 3.13. This modification improved the device performance mainly short circuit current

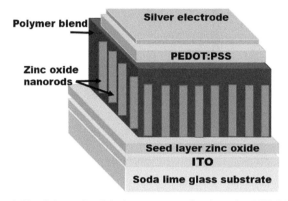

Figure 3.13 Schematic of device structure after inserting PEDOT:PSS.

Table 3.1 Comparison of ZnO and ZnO:Eu nanorods.

Sample	Sheet resistance	Photo response
ZnO nanorod	60 M$\Omega/\square$	0.1 A/W
ZnO:Eu nanorod	2 M$\Omega/\square$	4.5 A/W

density from 0.16 mA/cm^2 to 3.47 mA/cm^2 yielded an efficiency of 0.35 %. These type of modification in device structure and enhancing device perfor- mance is also explained elsewhere [48].

To further improve the collective behaviour of the heterostructure, ZnO nanorod arrays were doped with Europium (Eu) without disturbing its mor- phology and alignment. Effect of Eu doping on the properties of ZnO nanorods were analysed by measuring its resistance and photo response and is tabulated in Table 3.1. The table revealed that the sheet resistance of the sample reduces by Eu doping whereas photo response increases. Both were good for an electron transport layer.

In the device, ZnO nanorods were replaced by ZnO:Eu nanorods. This resulted in further enhancement of short circuit current density from 3.47 to

Table 3.2 PV parameters of the fabricated solar cells using nanorods.

Device structure	Device name	V_{oc} (mV)	J_{sc} (mA/cm^2)	FF (%)	η (%)
ITO/ZnO:Al (seed layer)/ZnO (nanorods)/P3HT:PCBM/Ag	D1	173	0.16	36	0.01
ITO/ZnO:Al(seed layer)/ZnO (nanorods)/P3HT:PCBM/PEDOT:PSS/Ag	D2	299	3.47	34	0.35
ITO/ZnO:Al (seed layer)/ZnO:Eu (nanorods)/P3HT:PCBM/PEDOT:PSS/Ag	D3	248	4.90	44	0.44

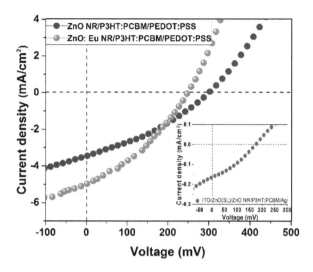

Figure 3.14 J-V characteristics of the devices corresponding to D1, D2 and D3 in Table 3.2.

4.90 mA/cm^2 and hence the efficiency from 0.35 % to 0.44 %. PV parameters are tabulated in Table 3.2 [48] and its J-V characteristics are shown in Figure 3.14.

3.3 Tangled Nano- and Micro-Root Structure for Photovoltaic Application

Here, nanorods are replaced by ZnO thin film with tangled root structure and its effect on device performance were analysed. Tangled root structure was developed using CSP technique by varying its basic film deposition parameters (substrate temperature and spray rate). Initially substrate temperature was varied and optimised as 250 °C and then spray rate was varied and optimised as 9 ml/min. Detail description about the growth of tangled root structure is described elsewhere [56]. Hence developed ZnO was a combination of roots with diameters in both nano and micro size and SEM image of the sample is shown in Figure 3.15.

Solar cells were fabricated with this thin film of ZnO having intermixed nano and micro roots (ZnO-TR). Device structure was ITO/ZnO-TR/P3HT:PCBM/Ag. When the devices were characterised, most of them get shorted and only a few showed p-n junction. On illumination they exhibited only slight photo activity. Shorting of the device might be due to the porous region or gap between these root structures which made the organic layer to have a direct contact with ITO. Here active layer was polymer-PCBM blend.

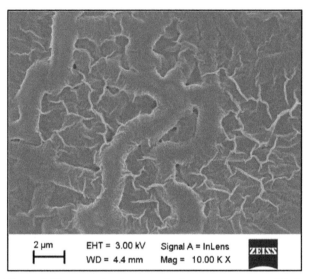

Figure 3.15   Intermixed nano and micro root structure of ZnO.

If polymer or PCBM made direct contact between top and bottom electrodes, device got shorted. Hence, a layer of thin film was required between ITO and ZnO-TR. Highly uniform ZnO thin film was chosen for this and CSP technique was employed for its deposition. Spray parameters were varied and optimised for the deposition of uniform ZnO layer. Spray conditions were substrate temperature 450 °C, spray rate 3 ml/min, carrier gas pressure 20 psi, molarity of precursor solution was 0.1 M. Samples deposited at this condition yielded a uniform film with lower surface roughness (~7 nm). But for device fabrication (schematic is in Figure 3.16) thickness is the major factor that determines the device performance. Hence, optimum thickness had to be determined by varying the thickness of uniform ZnO layer.

Solution volume was adjusted, to achieve the required thickness for the thin film, as 15 ml, 25 ml, 35 ml, 45 ml and 55 ml. Figure 3.17 depicts the PV parameters obtained for the devices at different thickness. This variation mainly affected the short circuit current density of the device. In other terms, increasing thickness of uniform ZnO layer from 80 nm to 128 nm enhances the charge carrier collection, but further increase in thickness reduces the carrier collection. This might be due to charge carrier recombination. But on analysing the efficiency of the devices at uniform ZnO layer thickness of 115 nm and 128 nm, it was similar, that is, device performances were better at uniform ZnO layer thickness ~120 nm. Best efficiency for the device obtained from this study was 1.39 %.

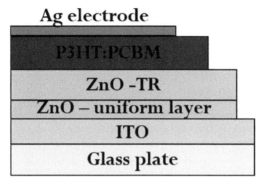

Figure 3.16   Schematic representation of device structure.

Volume of ZnO precursor solution	15 ml	25 ml	35 ml	45 ml	55 ml
ZnO sample thickness	80 nm	100 nm	115 nm	128 nm	145 nm

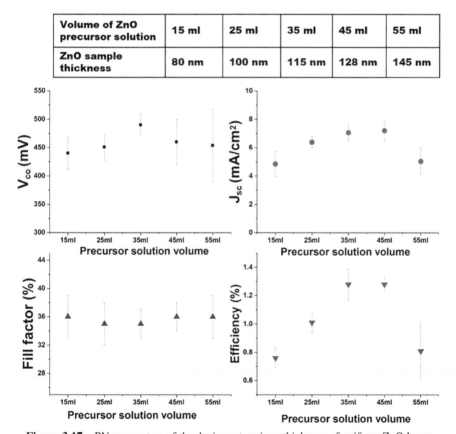

Figure 3.17   PV parameters of the devices at various thickness of uniform ZnO layer.

3.4 Conclusion

Hybrid thin film solar cells utilising ZnO nanomaterials were fabricated. Conditions for the deposition of ZnO nanorods, device fabrication using ZnO nanorods and its modification using PEDOT:PSS layer and Eu doping were discussed. Highest efficiency obtained for those devices were ~0.44 %. Another type of nanomaterial incorporation was done by CSP technique itself by slightly adjusting its parameters. An intermixed nano and micro sized (diameter) root structure was utilised for device fabrication and it yielded an efficiency of ~1.4 %. Comparatively device performance of ZnO nanorods was lower. This might be due to the improper infiltration of active layer into the gaps of ZnO nanorods. Adjusting the diameter of ZnO nanorods and density of nanorod array might enhance the device performance. Moreover, instead of intermixed nano and micro root, if we can reduce the diameter of all roots to nanoscale, then the device may perform much better due to the enhancement in surface area. A point to be noted is that for the device fabrication discussed above, deposition of ZnO layers and active layers were done under open air atmosphere without using any glove box.

References

[1] H. Wei, E. Wangg, "Nanomaterials with enzyme-like characteristics (nanozymes): next-generation artificial enzymes". Chem. Soc. Rev., **42** (2013) 6060.

[2] A . Juzgado, A. Solda, A. Ostric, A. Criado, G. Valenti, S. Rapino, G. Conti, G. Fracasso, F. Paolucci, M. Prato, "Highly sensitive electro-chemiluminescence detection of a prostate cancer biomarker". J. Mater. Chem. B., **5** (2017) 6681.

[3] J-P Kaiser, L. Diener, P. Wick, "Nanoparticles in paints: A new strategy to protect façades and surfaces?", Journal of Physics: Conference Series, **429** (2013) 012036.

[4] M. Anis, G. AlTaher, W. Sarhan, M. Elsemary (2017). Nanovate. Springer. p. 105.

[5] J. Lee, S. Mahendra, P. J. J. Alvarez, "Nanomaterials in the construction industry: A review of their applications and environmental health and safety considerations", ACS Nano., **4(7)** (2010) 3580–359.

[6] K. Gajanan, S.N.Tijare, "Application of Nano Materials", Materials Today: Proceedings, **5** (2018) 1093–1096

[7] S.G. Ullattil, S.B. Narendranath, S.C. Pillai, P. Periyat, "Black TiO$_2$ Nanomaterials: AReview of Recent Advances", Chem. Eng. J., **343** (2018) 708–736.

[8] C. Buzea, I. Pcheco, K. Robbie, "Nanomaterials and Nanoparticles: Sources and Toxicity". Biointerphases., **2(4)** (2007) MR17–MR7.

[9] B. Kumar and S. Kim, "Energy harvesting based on semiconducting piezoelectric ZnO nanostructures", Nano Energy, **1** (2012) 342.

[10] M. Salem, S. Akir, I. Massoudi, Y. Litaiem, M. Gaidi, K. Khirouni, "Photoelectrochemical and optical properties tuning of graphene-ZnO nanocomposites", Journal of Alloys and Compounds, **767** (2018) 982–987.

[11] Z. Xie, S Sun,W. Wang, L. Qin, Y. Yan, R. Hou, G.G. Qin, "Simulation study on improving efficiencies of perovskite solar cell: Introducing nano textures on it", Opt. Commun., **410** (2018) 117–122.

[12] B. A. Kumar, G.Sivasankar, B.S. Kumar, T.Sundarapandy, M.Kottaisamy, "Development of Nano-composite Coating for Silicon Solar Cell Efficiency Improvement", Materials Today: Proceedings, **5** (2018) 1759–1765.

[13] D. Z. Dimitrov, Chen-Hsun Du, "Crystalline silicon solar cells with micro/nano texture", Applied Surface Science **266** (2013) 1–4.

[14] Md. A. Mahmud, N. K. Elumalai, B. Pal, R. Jose, M. B. Upama, D. Wang, V. R.Gonçales, C. Xu, F. Haque, A. Uddin, "Electrospun 3D composite nano-flowers for high performance triple-cation perovskite solar cells", Electrochimica Acta, **289** (2018) 459–473.

[15] V. Khoshdel, M. Joodaki, M. Shokooh-Saremi, "UV and IR cut-off filters based on plasmonic crossed-shaped nano-antennas for solar cell applications", Opt. Commun.**433** (2019) 275-282.

[16] A. Timoumi, S. N. Alamri, H. Alamri, "The development of TiO2-graphene oxide nano composite thin films for solar cells", Results Phys., **11** (2018) 46-51.

[17] Y. Bai, I. Mora-Sero, F. D. Angelis, J. Bisquert, P. Wang, "Titanium Dioxide Nanomaterials for Photovoltaic Applications", Chem. Rev., **114 (19)**(2014) 10095–10130.

[18] A. B. DjuriŽiæ, X. Liu, Y. H. Leung, "Zinc oxide films and nanomaterials for photovoltaic applications", Phys. Status Solidi RRL **8** (2014) 123–132 (2014)

[19] L. Li, T. Zhai, Y. Bando, and D. Golberg, "Recent progress of one-dimensional ZnO nanostructured solar cells", Nano Energy, **1** (2012) 91.

[20] S. H. Ko, D. Lee, H. W. Kang, K. H. Nam, J. Y. Yeo, S. J. Hong, C. P. Grigoropoulos,H. J. Sung, "Nanoforest of hydrothermally grown hierarchical ZnO nanowires for a high efficiency dye-sensitized solar cell", Nano Letters, **11** (2011) 666.

[21] L. Campo et al., "Electrochemically grown ZnO nanorod arrays decorated with CDS quantum dots by using a spin-coating assisted successive-ionic-layer-adsorption and reaction method for solar cell applications", ECS J. Solid State Sci.Technol., **2(9)** (2013) Q151-Q158.

[22] W.J.E. Beek, M. M. Wienk, R. A. J. Janssen, "Hybrid solar cells from regioregular polythiophene and ZnO nanoparticles", Adv. Funct. Mat., **16** (2006) 1112–1116.

[23] Y. Sung, F. Hsu, Y. Chen, "Improved charge transport in inverted polymer solar cells using surface engineered ZnO-nanorod array as an electron transport layer", Sol. Energ. Mat. Sol. Cells, **125** (2014) 239–247.

[24] N. Sekine, C. Chou, W. Kwan, Y. Yang, "ZnO nano-ridge structure and its application in inverted polymer solar cell," Org. Electron., **10** (2009) 1473–1477.

[25] Y. Ho, P. Ho, H. Lee, S. Chang, "Enhancing performance of inverted polymer solar cells using two-growth ZnO nanorod", Sol. Energ. Mat. Sol. Cells, **132** (2015) 570–577.

[26] Z. Hu, J. Zhang, Y. Liu, Z. Hao, X. Zhang, Y. Zhao, "Influence of ZnO interlayer on the performance of inverted organic photovoltaic device", Sol. Energ. Mat. Sol. Cells, **95** (2011) 2126–2130.

[27] M. Tanveer, A. Habib, and M. Bilal, "Improved efficiency of organic / inorganic photovoltaic devices by electrospun ZnO nanofibers", Mat. Sci. Eng. B, **177** (2012) 1144–1148.

[28] K. C. Choi, E. J. Lee, Y. K. Baek, D. C. Lim, Y. C. Kang, Y. D. Kim, K. H. Kim, J. P. Kim, Y. K. Kim, "Morphologically controlled ZnO nanostructures as electron transport materials in polymer-based organic solar cells", Electrochimica Acta, **180** (2015) 435–441.

[29] F. Tong, K. Kim, D. Martinez, R. Thapa, A. Ahyi, J. Williams, D. J. Kim, S. Lee, E. Lim, K. K. Lee, M. Park, "Flexible organic/inorganic hybrid solar cells based on conjugated polymer and ZnO nanorod array", Semicond. Sci. Tech., **27** (2012) 105005 (1–5).

[30] B. Gholamkhass, N. M. Kiasari, P. Servati, "An efficient inverted organic solar cell with improved ZnO and gold contact layers", Org. Electron., **13** (2012) 945–953.

[31] S. K. Hau, H. L. Yip, N. S. Baek, J. Zou, K. O'Malley, A. K. Y. Jen, "Air-stable inverted flexible polymer solar cells using zinc oxide nanoparticles as an electron selective layer", Appl. Phys. Lett., **92** (2008) 1–4.

[32] S. H. Oh, S. J. Heo, J. S. Yang, H. J. Kim, "Effects of ZnO nanoparticles on P3HT:PCBM organic solar cells with DMF-modulated PEDOT:PSS buffer layers", ACS Appl. Mater. Interfaces, **5** (2013) 11530–11534.

[33] T. H. Lee, H. J. Sue, X. Cheng, "ZnO and conjugated polymer bulk heterojunction solar cells containing ZnO nanorod photoanode", Nanotechnology, **22** (2011) 285401 (1-6).

[34] C. Chou, J. Huang, C. Wu, C. Lee, C. Lin, "Lengthening the polymer solidification time to improve the performance of polymer/ZnO nanorod hybrid solar cells", Sol. Energ. Mater. Sol. Cells, **93** (2009) 1608–1612.

[35] N. Sekine, C. H. Chou, W. L. Kwan, Y. Yang, "ZnO nano-ridge structure and its application in inverted polymer solar cell", Org. Electron., **10** (2009) 1473–1477.

[36] Y. M. Sung, F. C. Hsu, C. T. Chen, W. F. Su, Y. F. Chen, "Enhanced photocurrent and stability of inverted polymer/ZnO-nanorod solar cells by 3-hydroxyflavone additive", Sol. Energ. Mater. Sol. Cells, **98** (2012) 103–109.

[37] C. Han Y. Chen, L. Chen, L. Qian[1], Z. Yang W. Xue, T. Zang, Y. Yand, W. Cao, "Enhanced performance of inverted polymer solar cells by combining ZnO nanoparticles and poly[(9,9-bis(3'-(N,N-dimethylamino)propyl)-2,7-fluorene)-alt-2,7-(9,9-dioctyfluorene)] as electron transport layer", ACS Appl. Mater. Interfaces, **8(5)** (2016) 3301–7.

[38] Y. Y. Lin, Y. Y. Lee, L. Chang, J. J. Wu, C. W. Chen, "The influence of interface modifier on the performance of nanostructured ZnO/polymer hybrid solar cells", Appl. Phys. Lett., **94** (2009) 10–13.

[39] M.V. Santhosh, D.R. Deepu, C. Sudha Kartha, K. Rajeev Kumar, K.P. Vijayakumar, "All sprayed ITO-free CuInS$_2$/In$_2$S$_3$ solar cells",J. Sol. Energy, **108** (2014) 508–514.

[40] K. S. Ranjith, R. Geethu, K. P. Vijayakumar, R. T. Rajendrakumar, "Control of interconnected ZnO nanowires to vertically aligned ZnO nanorod arrays by tailoring the underlying spray deposited ZnO seed layer", Mater. Res. Bull., **60** (2014) 584–588.

[41] R. R. Philip, B. Pradeep, "Structural analysis and optical and electrical characterizations of the ordered defect compound CuIn$_5$Se$_8$", Semicond. Sci. Tech., **18** (2003) 768–773.

[42] K.S. Ranjith et al., Alignment, morphology and defect control of vertically aligned ZnO NR array: Competition between "Surfactant" and "Stabilizer" roles of the amine species and its photocatalytic properties, Cryst. Growth Des., **14** (2014) 2873–287.

[43] K.S. Ranjith et al., Optimisation on the growth and alignment of ZnO NRs, Adv. Mater. Res, **584** (2012) 319–323.

[44] K. S. Ranjith, "Defect controlled, solution grown, vertically aligned ZnO NR arrays: Study on improved photocatalytic and photovoltaic properties, (2015) Bharathiar University, Coimbatore.

[45] P. Jiang, M.J. McFarland, "Large-scale fabrication of wafer-size colloidal crystals, macroporous polymers and nanocomposites by spin-coating, J. Amer. Chem. Soc., **126(42)** (2004) 13778–13786.

[46] https://www.ossila.com/pages/spin-coating#dynamic-dispense-spin-coating-technique

[47] http://www.utdallas.edu/~rar011300/CEEspinner/SpinTheory.pdf

[48] Geethu R., "Hybrid thin film solar cell fabrication using sprayed electron collective layer and spin or spray coated organic layer", PhD Thesis (May 2018-submitted), Cochin University of Science and Technology, Cochin.

[49] D.A. Neamen(1992)"Semiconducting physics and devices – Basic principles", Irwin Inc., New York

[50] J. Nelson, "The physics of solar cells", Imperial College Press, UK.

[51] A. L. Fahrenbruch, R. H. Bube(1983), "Fundamentals of solar cells", First ed.,Academic Press, USA.

[52] http://pveducation.org/pvcdrom/short-circuit-current

[53] http://pveducation.org/pvcdrom/solar-cell-operation/fill-factor

[54] T.V. Vimalkumar, N. Poornima, K.B. Jinesh, C. Sudha Kartha, K.P. Vijayakumar, "On single doping and co-doping of spray pyrolysed ZnO films: Structural, electrical and optical characterisation", Appl. Surf. Sci., **257** (2011) 8334.

[55] T.V. Vimalkumar, N. Poornima, C. Sudha Kartha, K.P. Vijayakumar, T. Abe,Y. Kashiwaba, "Enhancement of electrical conductivity in sprayed ZnO thin film through zero-energy process", Physica B-Condensed Matter, **405** (2010) 4957.

[56] R. Geethu, C. Sudha Kartha, K.P. Vijayakumar, "Improving the performance of ITO/ZnO/P3HT:PCBM/Ag solar cells by tuning the surface roughness of sprayed ZnO", J. Sol. Energy, **120** (2015) 65–71.

Part II

Nanostructured Polymer Composites

4

Effect of Nanosilica Concentration on the Mechanical, Viscoelastic and Morphological Properties of Polypropylene/Styrene– Ethylene/Butylene–Styrene Blend

Harekrishna Panigrahi[1,*], Smrutirekha Mishra[2], Avinash Nath Tiwari[1] and M. P. Singh[3]

[1]Department of Mechanical Engineering, Jagannath University, Jaipur, Rajasthan, India
[2]School of Chemical Technology, Kalinga Institute of Industrial Technology (KIIT), Bhubaneswar, Odisha, India
[3]Department of Mechanical Engineering, JECRC University, Jaipur, Rajasthan, India
E-mail: harekrishnapanigrahi91@gmail.com
*Corresponding Author

This work attributes on the effects of nanosilica of size up to 60–80 nm as an inorganic nanofiller for the enhancement of mechanical properties of PP/SEBS blends. Different compositions of PP/SEBS/Si blend nanocomposites are formulated and prepared by melt mixing method. Affluent preparation of the blends is shown in the morphological studies of the polymeric blends, and the distribution of nanosilica in the matrix was deliberated using high-resolution transmission electron microscopy (HRTEM). Tensile tests for 10 wt% nanosilica manifested superlative increase of approximately 35% higher than the other loading percentage of nanosilica in the prepared blends. In subsistence to tensile tests performed dynamic mechanical analysis (DMA), studies also revealed a virtuous storage modulus at 30°C for 10 wt% loading of nanosilica arraying admirable viscoelastic properties of the prepared blends.

Graphical abstract

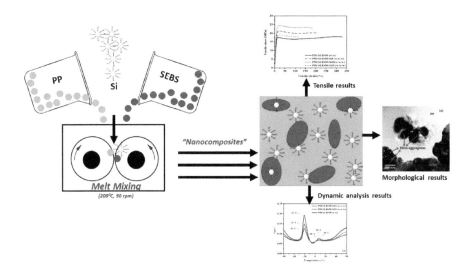

4.1 Introduction

Polymer blends can be narrated as the bodily aggregate of two or greater polymers [1-2]. Blending of polymers furnishes novel materials that may amalgamate the properties of polymers that are not attainable with individual component [3]. The conventional lucrative polymer blends may be plastic-plastic, plastic-rubber or rubber-rubber [4]. Application of these eminent blends precedence to a reduction in the amount of the more costly material vital, and/ or to a development in the properties [4]. Overviewing divergent types polymeric blends, plastic–rubber blends habitually occupied their position in countless commercial applications.

Polypropylene (PP) is considered as one of the topmost assuring substitutes for engineering substances due to its splendid processability, rigidity, thermal stability, resistance to oil, recyclability and relatively low cost [5]. To improve PP competitiveness in engineering applications, a simultaneously boost in stiffness and toughness is essential. Toughness, a very critical mechanical property that reflects the material potential to absorb the impact energy, may be drastically better by using the incorporation of a dispersed rubbery section. However, the impact power of PP can be amplified by reckoning a bit of rubbery material such as ethylene–propylene–diene monomer, butadiene–styrene–acrylonitrile terpolymer, acrylonitrile–butadiene rubber, styrene–butadiene–styrene copolymers and ethylene–octane copolymers [5-10]. On this behalf, countless research related to PP–poly [styrene-(ethylene-co-butylene)-styrene] triblock copolymer (SEBS) blends have been studied [11-18]. A study by Gupta and Purwar showed the mechanical, dynamic mechanical and rheological properties of PP blends as a function of SEBS percentage materials as well as crystallisation behaviour of PP in the SEBS phases. A fine brittle–tough transition became discovered at SEBS contents of approximately 15 vol% [15-18]. Likewise Stricker et al. have studied the importance of particle size of rubber on mechanical properties of PP-SEBS blend [12]. Yield stress and stiffness of PP and PP blends boost with increasing PP molecular weight in the equal extent, and are consequently most effective depending on matrix characteristics and not on dispersed SEBS particle size [12].

Traditionally thermoplastic polyolefin TPO blends are known to have beneath mechanical strength (tensile strength) and stiffness analogously lower than of pure PP due to the addition of elastomer. Thus, to perpetuate the properties of such blends and nanocomposites (mechanical strength and stiffness) inorganic nanofillers have played an important role [11]. The mantle of these inorganic nanofiller on the physico-mechanical properties of nanocomposite heavily depends on its surface properties, shape and aggregating size. One important parameter toughening by inorganic nanofiller is the dispersion of these nanofiller in the matrix [11]. To best of our knowledge, very few literatures have reported the preparation and different properties of PP/SEBS-based nanocomposites [5, 19-22]. Vuluga et al. have studied the effect of SEBS on thermal, mechanical and morphological properties of PP/organoclay-based nanocomposites [5] reporting that huge enhancement in the impact strength (nearly 22 times) became acquired in the case of SEBS-containing nanocomposites in contrast with the composite without

SEBS [5]. Similarly, Majid et al. examined the effect of styrene–butadiene–styrene content on morphology, melting, crystallinity, dynamic mechanical properties and relaxation processes of polypropylene/poly(styrene-co-acrylonitrile)/styrene–butadiene–styrene blends. Styrene–butadiene–styrene reduced the average size of dispersed particles and generated complex aggregates in the matrix. Morphology development examined by dynamic mechanical thermal analysis showed increased damping of poly (styrene-co-acrylonitrile) domains at high styrene–butadiene–styrene contents. All blends showed reduced crystallinity and melting point compared with neat polypropylene [19]. Noah et al. studied the influences of incorporating compatibilisers E-EA-MAH, E-MA-GMA, E-AM, SEBS KRATON G or PP-g-MAH on the thermal properties of mixed (polypropylene/ethylene propylene rubber)/acrylonitrile butadiene styrene (PP/EPR)/ABS. DSC investigations have revealed that the incorporation of 5% of ABS in the copolymer (PP/EPR) does not fundamentally affect the thermal properties of the basic copolymer; additionally, the addition of 1.5% of each of the compatibilisers in the basic mixture does not significantly alter the crystallisation temperature values and the melting of the -P- sequences [20]. Likewise, Sanporean et al. have studied the toughness and stiffness properties of PP/SEBS/organoclay-based nanocomposite [21]. They have shown that the nanocomposites provide improved thermal stability, 300% improvement in impact strength and small decrease in tensile strength [21]. For more example, Panaitescu et al. have studied the morphological properties of PP/SEBS/nanosilica-based nanocomposites [22]. They have found that the boost of all tensile characteristics is located in PP/nanosilica composites without SEBS and contains 5% SEBS and a large plastic deformation inside the nanocomposite with 10% SEBS [22].

From the literature, it is evident that very less work has been reported for the enhancements of mechanical properties by adding silica nanoparticles on PP/SEBS blends. Consequently, this chapter specialises in the effect of different ratio of nanosilica particles on the mechanical properties of 50PP/50SEBS blend. In this work, nanocomposites based on PP/SEBS/nanosilica were prepared by melt mixing. Different concentrations of nanosilica were used as fillers in this study in order to explore their influences on mechanical properties and contribute to the enhancement of already existent results by correlating mechanical-dynamic mechanical and morphological properties.

Table 4.1 Mixing formulations.

Sl No	Composition of sample	Polypropylene (Wt %)	SEBS (Wt %)	Nanosilica (Wt %)
1	P50/SEBS50 (w/w)	50	50	-
2	P50/SEBS50/Si5 (w/w/w)	50	50	5
3	P50/SEBS50/Si10 (w/w/w)	50	50	10
4	P50/SEBS50/Si15 (w/w/w)	50	50	15

4.2 Experimental Part

4.2.1 Materials Used

Polypropylene (PP) pellets (grade-1110MAS Homopolymer injection grade) of melt flow index MFI=11 g/10 min, melting point 170°C and density= 0.9 g/cc, supplied by IOCL was used for blend and nanocomposites preparation. Styrene-ethylene/butylene-styrene block copolymer (SEBS) (grade - G1650) with 29.2% polystyrene, MFI = 4 g/10 min (230°C per 5 kg) supplied by Kraton Polymers USA. Precipitated nanosilica particles were procured from Madhu Pvt. Ltd., Bhavnagar, Gujarat (grade-MS140) in powder form.

4.2.2 Preparation of Nanocomposites

PP and SEBS were melt-mixed in a Haake Rheocord internal mixer of batch weight 50 g with counter rotating roller type rotor to form PP/SEBS blend. The mixing was performing at 200°C temperature and 90 rpm rotor speed for 6 min. In order to make PP/SEBS/nanosilica nanocomposites SEBS and nanosilica were mixed first at 80°C, 90 rpm for 5 min. After that PP and premixed (SEBS + nanosilica) were melt-mixed at previous condition for 6 min to make PP/SEBS nanocomposites. After complete mixing in Haake Rheocord internal mixer, the resulting blends were removed and then compression moulded (using hydraulic press, Moore & Son Ltd., UK) at 200°C, 5 MPa pressure for 4 min and finally cooled.

4.2.3 Characterisation

4.2.3.1 Mechanical properties

Tensile test of the dumbbell shaped specimens was carried out on a universal testing machine (Instron 3382) according to ASTM D638. The crosshead speed for testing the tensile samples was set at 200 mm/min at room temperature using a load cell of 10 kN. The result reported here was the average of four samples from the same batch.

4.2.3.2 Viscoelastic properties

The viscoelastic properties of the blends and nanocomposites were studied by dynamic mechanical analyser (DMA) of TA Instruments model Q800) in tension mode. Specimens of dimension ($16\times9.8\times1$) mm were analysed in the temperature range of -90 to $90°C$ at a controlled heating rate of $2°C/min$ at constant sinusoidal frequency of 1 Hz and at strain amplitude of 10 μm.

4.2.3.3 Morphological properties

Bulk morphological properties of polymer nanocomposite specimens were studied using high-resolution transmission electron microscopy (HRTEM) (model JEOL 1210, Tokyo, Japan) operated at 200 kV accelerated voltage. The cryosections were prepared by ultra-cryomicrotomy at $-90°C$ using Leica Ultracut UCT, Leica Microsystems (Vienna, Austria).

4.3 Results and Discussion

4.3.1 Tensile Stress-Strain Studies

Figure 4.1 represents tensile stress versus tensile strain graphs of PP/SEBS blend and PP/SEBS nanocomposite samples. Table 4.2 represents the mechanical properties for the same and pristine PP. Pristine PP exhibited higher tensile strength and lower elongation at break in comparison to P50/SEBS50 blend and their nanocomposites (Table 4.2). The incorporation of SEBS into PP results in a huge increase of PP ductility as shown in Figure 4.1. It was observed that the P50/SEBS50 blend reflected higher elongation at break (330%), lower tensile strength (18.1 MPa). In order to improve the tensile strength, nanosilica was added as filler at different concentrations (i.e. 5 wt%, 10 wt% and 15 wt%) in P50/SEBS50 blend to make nanocomposites. The tensile strength and modulus at 100% elongation values show a remarkable increment with an increase in nanosilica concentration of up to 10 wt% in the P50/SEBS50 blend. But tensile strength and modulus at 100% elongation values show a remarkable decrement with an increase in nanosilica concentration at 15 wt% in the P50/SEBS50 blend.

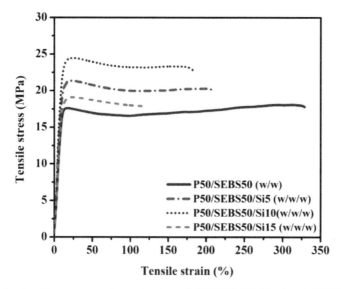

Figure 4.1 Tensile stress–tensile strain plots of P50/SEBS50 blend and P50/SEBS50/Si5, P50/SEBS50/Si10, P50/SEBS50/Si15 nanocomposites.

Table 4.2 Tensile stress–tensile strain properties of neat PP, P50/SEBS50 blend and P50/SEBS50/Si5, P50/SEBS50/Si10, P50/SEBS50/Si15 nanocomposites.

Sl. no	Sample details	Tensile strength (MPa)	Elongation at break (%)	Modulus @ 100% elongation (MPa)
1	Neat PP	34 ±1	19±5	-
2	P50/SEBS50	18.1±0.6	330±8	16.6±0.4
3	P50/SEBS50/Si5	21.3±0.5	210±7	20±0.5
4	P50/SEBS50/Si10	24.4±0.7	190±8	23.5±0.6
5	P50/SEBS50/Si15	19.2±1	120±6	18±0.8

Gazing to the mechanical properties of P50/SEBS50/Si15 (w/w/w), it seems to decrease due to more filler–filler interaction leading to high agglomeration of the nanoparticles, which is clearly perceived through HRTEM. Contrasting it with P50/SEBS50/Si10 nanocomposite, it can be explicitly seen and assumed that the dispersion of silica is exemplary which is hence proved by HRTEM [25].

For example, tensile strength value of P50/SEBS50 blend rises by 18%, 35% and 6% for 5 wt%, 10 wt% and 15 wt% loading of nanosilica, respectively. Elongation at break values constantly decreases with increase in the percentage of nanosilica in P50/SEBS50 blend (Table 4.2). Modulus at 100% elongation value of P50/SEBS50 blend increases by 20%, 41% and 8% for 5 wt%, 10 wt% and 15 wt% loading of nanosilica, respectively. It is worth mentioning here that well-dispersed nanosilica particles in PP and SEBS phases develop in a strong and elastomeric material at an optimum nanosilica concentration (10 wt %). The dispersion of nanosilica particles in PP and SEBS matrix at numerous concentrations has been explained concisely inside the morphological section. Storage modulus values at 30°C derived from DMA are matching with tensile stress-tensile strain plot. These effects are explained in the upcoming section.

4.3.2 Dynamic Mechanical Analysis (DMA)

Figure 4.2a represents the loss tangent versus temperature graphs of PP and SEBS blends and their nanocomposites over the temperature range of -90°C to 90°C. From Table 4.3, it has been noted that within the above temperature range neat PP represented a single T_g at 13°C, whereas neat SEBS represented T_g at -29°C due to the relaxation of soft rubbery ethylene-butylene segments [13]. In PP and SEBS blend (50/50) two distinct peaks represented the T_g of SEBS and PP. The presence two distinct tan δ peaks strongly represent that two separate phase exist in PP-SEBS blend.

From Figure 4.2a, it has been noted that P50/SEBS50 blend shows two loss tangent peaks at -31°C and 10°C which represents the T_g values of SEBS and PP, respectively. It should be pointed here that the addition of (10 and 15 wt%) nanosilica particles into P50/SEBS50 blend does not change the T_g values of PP and SEBS. However, incorporating the nanosilica particles into P50/SEBS50 blends significantly reduces the loss tangent peak height corresponding to PP and SEBS phase. The decrements in loss tangent peak height values are more prominent at lower concentration (10 wt %) of nanosilica as compared to higher concentration (15 wt %) of nanosilica (Table 4.3). Some authors have already reported that the decrement of loss tangent peak height of any polymer by the incorporation of any filler can provide reinforcing effect of filler in the polymer matrix [23-24]. Here, the decrement in loss tangent peak height values can be attributed to the more reinforcing effect of nanosilica particles in P50/SEBS50/Si10 nanocomposites.

Storage modulus (E') values at 30°C of neat PP and SEBS, their blends and nanocomposites are presented in Table 4.3. Figure 4.2b represents the storage modulus (E') versus temperature graphs of PP and SEBS blends and their nanocomposites over the temperature range of -90 to 90°C. It has

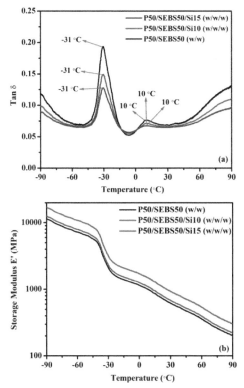

Figure 4.2 (a) Tan δ versus temperature graphs of P50/SEBS50 blend and P50/SEBS50/Si10, P50/SEBS50/Si15 nanocomposites. (b) Storage modulus (E') MPa versus temperature graphs of P50/SEBS50 blend and P50/SEBS50/Si10, P50/SEBS50/Si15 nanocomposites.

Table 4.3 T_g, Tan δ peak height (Tan δ_{max}) and elastic modulus (E') values for neat PP, P50/SEBS50 blend and P50/SEBS50/Si10, P50/SEBS50/Si15 nanocomposites.

Sl. no	Sample	T_{gru} [a] (°C)	T_{gpl} [b] (°C)	Tan δ_{max} [c]	E' (MPa) at 30°C
1	Neat PP	-	13	-	607
2	Neat SEBS	-29	-	1.2	3.2
3	P50/SEBS50	-31	10	0.19	290
4	P50/SEBS50/Si10	-31	10	0.12	423
5	P50/SEBS50/Si15	-31	10	0.14	312

[a] Glass transition temperature of the rubber phase.
[b] Glass transition temperature of the plastic phase.
[c] Tan δ value at the maximum peak position of the rubber phase

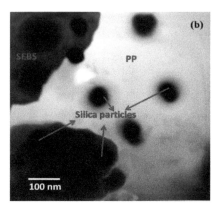

Figure 4.3   (a) HRTEM photomicrograph of P50/SEBS50/Si15 nanocomposite. (b) HRTEM photomicrograph of P50/SEBS50/Si10 nanocomposite.

been observed that the storage modulus value (at 30°C) of P50/SEBS50 blend significantly increases with the addition of 10 wt% of nanosilica. By incorporating 15 wt% of nanosilica in P50/SEBS50 blend the modulus (at 30°C) reduces. The reduction of storage modulus is due to more filler–filler interaction, which will be discussed in the morphological section.

The presence of two distinct tan δ peaks strongly represents that two separate phases exist in PP-SEBS blend which is proved by HRTEM micrographs. On the other hand, incorporation of 15% of nanosilica particles does not significantly improve the tensile properties (maximum tensile strength and modulus at 100% elongation) and DMA storage modulus values (at 30°C) due to more silica aggregates (more filler–filler interaction).

4.3.3 High-resolution Transmission Electron Microscopy (HRTEM)

Figure 4.3b represents the HRTEM photomicrograph of P50/SEBS50/Si10 nanocomposite. From this micrograph it can clearly observed the non-aggregated structure of nanosilica particles. The micrograph showed that P50/SEBS50/Si10 nanocomposite is developed and the nanosilica particles are well dispersed inside the polymer matrix. Due to more polymer–filler interaction, the nanosilica particles provide more reinforcing effect and improve its mechanical properties. This is matching with the results obtained from DMA and tensile stress–strain studies (Tables 4.2 and 4.3). Figure 4.3a represents the HRTEM photomicrograph of P50/SEBS50/Si15 nanocomposite. From this micrograph, the aggregated structure of nanosilica particles can be clearly observed. Due to more filler–filler interaction, the nanosilica particles do not provide reinforcing effect. This is matching with the results obtained from DMA and tensile stress–strain studies (Tables 4.2 and 4.3).

4.4 Conclusion

Effect of nanosilica on the mechanical, viscoelastic and morphological properties of PP/SEBS blend has been investigated. Significant improvement in the mechanical properties of the P50/SEBS50 blend has been observed by the incorporation of nanosilica. Tensile stress value of P50/SEBS50 blend containing 10 wt% of nanosilica is approximately 35% higher than the tensile stress value of P50/SEBS50 blend. The mechanical properties of P50/SEBS50 blend start to decrease when the nanosilica concentration increases from 10 to 15 wt%. The DMA storage modulus value (at 30°C) of P50/SEBS50 blend containing 10 wt% of nanosilica is approximately 46% higher than the DMA elastic modulus value (at 30 °C) of neat P50/SEBS50 blend. HRTEM shows the existence of non-aggregated structure of nanosilica morphology in P50/SEBS50 blend containing 10 wt% of nanosilica. However, at 15 wt% of nanosilica concentration, HRTEM confirms the existence of aggregated morphology of the nanosilica in P50/SEBS50 blend.

References

[1] Utracki L.A. Introduction to polymer blends. In: L.A. Utracki (Ed.), Polymer Blends Handbook, Kluwer academic publishers, Dordrecht, 2002, pp. 01-24.

[2] Dixon K.W. Polymerization and depolymerization. In: J. Brandrup, E.H. Immergut, E.A. Grulke (Eds.), Polymer Handbook, John Wiley and Sons, Inc., New York, 1999, pp. 01-30.

[3] Utracki L.A. and Favis B.D. Polymer alloys and blends. In: N.P. Cheremisinoff (Ed.), Handbook of Polymer Science and Technology, Marcel Dekker Inc., New York, 1989, pp. 121-202.

[4] Avramova N. Polymer additives: The miscibility of blends. In: Pritchard G. (Ed.) Plastics Additives. Polymer Science and Technology Series, Springer, Dordrecht. **vol 1**,1998, pp. 523-514

[5] Vuluga Z., Panaitescu D.M., Radovici C., Nicolae C. and DoinaIorga M. Effect of SEBS on morphology, thermal and mechanical properties of PP/organoclay nanocomposites, Polym. Bull., **69**, 2012, 1073–1091.

[6] Jain A.K., Nagpal A.K., Singhal R. and Gupta N.K. Effect of dynamic cross linking on impact strength and other mechanical properties of polypropylene/ethylene-propylene-diene rubber blends, J. Appl. Polym. Sci., **78**, 2000, 2089–2103.

[7] Tortorella N. and Beatty C.L. Morphology and mechanical properties of impact modified polypropylene blends, Polym. Eng. Sci., **48**, 2008, 2098–2110.

[8] Razavi-Nouri M., Naderi G., Parvin A. and Ghoreishy M.H.R. Thermal properties and morphology of isotactic polypropylene/acrylonitrile–butadiene rubber blends in the presence and absence of a nanoclay, J. Appl. Polym. Sci., **121**, 2011, 1365–1371.

[9] Ohlsson B. and Tornell B. Blends and interpenetrating polymer networks of polypropylene and poly styrene-block-poly (ethylene-stat-butylene)-block-polystyrene melt flow and injection molding properties, Polym. Eng. Sci., **38**, 1998, 108–118.

[10] Matsuda Y. and Hara M. Effect of the volume fraction of dispersed phase on toughness of injection molded polypropylene blended with SEBS, SEPS, and SEP, Polym. Eng. Sci., **45**, 2005, 1630–1638.

[11] Liao C.Z. and Tjong S.C. Effects of Carbon nanofibers on the fracture, mechanical, and thermal Properties of PP/SEBS-g-MA Blends, Polym. Eng. Sci., **51(5)**, 2011, 948-958.

[12] Stricker F., Thomann Y. and Lhaupt R. M. Influence of rubber particle size on mechanical properties of polypropylene–SEBS Blends, J. Appl. Polym. Sci., **68**, 1998, 1891–1901.

[13] Ahmad Z., Kumar K.D., Saroop M., Preschilla N., Biswas A., Bellare J.R. and Bhowmick A.K., Highly transparent thermoplastic elastomer from isotactic polypropylene and styrene/ethylene-butylene/styrene tri-block copolymer: structure-property correlations, Polym. Eng. Sci., **50(2)**, 2010, 332-341.

[14] Bassani A., Pessan L. A. and Hage E. Toughening of polypropylene with styrene/ethylene-butylene/styrene triblock copolymer: Effects of mixing condition and elastomer content, J. Appl. Polym. Sci., **82**, 2001, 2185–2193.

[15] Gupta A.K. and Purwar S. N. Dynamic mechanical and impact properties of PP/SEBS blend, J. Appl. Polym. Sci., **31**, 1986, 535-551.

[16] Gupta A.K. and Purwar S. N. Tensile yield behavior of PP/SEBS blends, J. Appl. Polym. Sci., **29**, 1984, 3513-3531.

[17] Gupta A. K. and Purwar S. N. Crystallization of PP in PP/SEBS Blends and its correlation with tensile properties, J. Appl. Polym. Sci., **29**, 1984, 1595-1609.

[18] Gupta A. K. and Purwar S. N. Melt Rheological properties of polypropylene/SEBS (styrene/ethylene-butylene/styrene triblock copolymer) blends, J. Appl. Polym. Sci., **29**, 1984, 1079-1093.

[19] Mazidi M.M., Hosseini F.S., Berahman R., Shekoohi K. and Basseri G., Phase morphology, thermal, thermomechanical and interfacial properties of PP/SAN/SBS blend systems, Polymer-Plastics Tech. and Engg., **56**, 2017, 254-267.

[20] Noah P. M., Ohandja L.M., Medjo R.E., Chabira S., Ebanda F. B., and Ondoua P. A., Study of thermal properties of mixed (PP/EPR)/ABS with five model compatibilizers, J. Engg., **5**, 2016, 1-9.

[21] Sanporean C.G., Vuluga Z., Radovici C., Panaitescu D. M., Iorga M., Christiansena J.C. and Moscac A. Polypropylene/organoclay/SEBS nanocomposites with toughness–stiffness properties, RSC Adv., **4**, 2014, 6573-6579.

[22] Panaitescu D. M., Vuluga Z., Radovici C. and Nicolae C., Morphological investigation of PP/nanosilica composites containing SEBS, Polym. Test., **31**, 2012, 355–365.

[23] Bandyopadhyay A., Thakur V., Pradhan S., and Bhowmick A.K. Nanoclay distribution and its influence on the mechanical properties of rubber blends, J. Appl. Polym. Sci., **115**, 2010, 1237–1246.

[24] Maiti S., De S. K., and Bhowmick A.K. Quantitative estimation of filler distribution in immiscible rubber blends by mechanical damping studies, Rub. Chem. Technol., **65**, 1992, pp. 293-302.

[25] Chi X., Cheng L., Liu W., Zhang W., and Li S., Characterization of polypropylene modified by blending elastomer and nano-silica, Materials, **11**, 2018, 2-13.

5

A Comparative Approach to Structural Heterogeneity of Polyaniline and Its ZnO Nanocomposites

Bhabhina Ninnora Meethal, P. C. Ajisha, Dharsana M. Vidyadharan, Jyothilakshmi V. Prakasan and Sindhu Swaminathan*

Department of Nanoscience and Technology, University of Calicut, Kerala, India
E-mail: sindhus@uoc.ac.in
*Corresponding Author

The polyaniline (PANI)-wrapped ZnO possesses an interface between organic semiconductor (polyaniline) and inorganic semiconductor (ZnO). Here highly crystalline polyaniline was synthesised through chemical approach. The obtained polyaniline has shown good crystalline behaviour compared to the commercially available emeraldine salt. The polyaniline–zinc oxide hybrid composites were also prepared through an easy one-pot facile method following two different protocols. In the first protocol, polyaniline was added during the formation of surface oxygen deficient ZnO (PNZ), while in the second protocol, zinc oxide nanotubes were added during the formation of polyaniline (ZNP). ZNP has emeraldine salt structure, whereas PNZ attains emeraldine base form. Various characterisation techniques were used to investigate the structural and morphological differences between these two nanocomposites. The adsorption ability of the synthesised polyaniline and the photocatalytic activity of PANI–ZnO nanocomposites have been checked by analysing the decolourisation of methylene blue dye solution. The reactive oxygen species (ROS) generated from the nanocomposites are quantified and compared. The photocatalytic mechanism responsible for the dye degradation process has been explained in detail. As a result of higher ROS generation, PNZ exhibits improved photocatalytic activity compared to ZNP.

5.1 Introduction

Conducting polymers, especially polyaniline, have been widely used for water treatment processes because of their ability to adsorb the dissolved pollutants effectively [1, 2]. The efforts of scientific community in developing hybrid nanocomposites of polymers with the aid of metal oxides augur well for implementing photocatalysis for mitigating pollution [3]. The myriad parameters that can be varied and have different material components which can be chosen to play a pivotal role in developing these composite materials. Hybrid nanocomposites either exhibit new properties or a mix-up of individual component characteristics. Wide band gap metal oxides, such as zinc oxide and their hybrid nanocomposites, with conducting polymers extensively improved their application in sensors [4], electronics [5], photovoltaics [6], batteries [7] and photocatalysis [1] due to their modified surface properties [8], band gaps [9], size [10] and shapes [11]. Surface modification of ZnO nanoparticles can be easily achieved through the incorporation of a polymer shell [4]. The synergetic and complimentary performance between conducting polymer and inorganic metal oxide makes a significant improvement in the properties of these hybrids. Over the past few decades, polyaniline is very much attractive with its intrinsic conduction for various catalytic processes including photocatalysis [12]. Crystalline polyaniline is found to be desirable for excellent photoelectrical applications with their exceptional conductivity [13]. The photocorrosion of ZnO can be inhibited through polyaniline coating [14]. Tuning the particle size and crystallinity of nanomaterials helps to increase their photocatalytic activity to a greater extend [15]. The polyaniline coating on ZnO surface stabilise the surface defects and modifies the photocatalytic activity [16].

Present work is an attempt to analyse the properties and photocatalytic activity of hybrid nanocomposites through polyaniline-assisted modification of ZnO nanotubes. Synthesis protocol for oxygen-deficient pristine zinc oxide nanotubes is reported elsewhere [17]. Addition of polyaniline during the formation of zinc oxide nanotubes resulted in a dark blue product, labelled as PNZ. Another nanocomposite with dark green colour is synthesised by adding the aforesaid zinc oxide during the formation of polyaniline (ZNP).

The strategy adopted here is a cost-effective, template-free and highly reproducible one-pot precipitation method. Out of various synthesis methods, simpler methods with cheap and minimum usage of precursors are always the major choice. Methods involving in situ nucleation and subsequent growth always get special attention because of their less time consumption and mild reaction condition. Synthesis of highly crystalline polyaniline (SP) through Fe^{3+}/H_2O_2 chemical oxidation route within a time frame of 2 h is reporting for the first time. Synthesis of nanocomposite has got considerable research attention due to their unique properties and formability. Specifically, active semiconductor-incorporated polymer nanocomposites have advantage of effective charge separation through the formation of an interface between them. Here the formation of nanocomposites was ensured from various systematic characterisation techniques. The structural, optical and morphological studies of these photocatalysts are discussed and compared in detail. The role of polyaniline-ZnO interface in the degradation of a model dye system is demonstrated. Moreover, the reactive oxygen species generated from both nanocomposites were estimated and quantified.

5.2 Experimental

5.2.1 Synthesis of Polyaniline and Its ZnO Nanocomposites

For the synthesis of polyaniline, 10 mL aniline was dissolved in 10 mL dilute HCl and stirred. After 5 minutes, a mixture of ferrous sulphate, hydrochloric acid and hydrogen peroxide (0.5 gm of $FeSO_4$ $7H_2O$, 15 mL of Con. HCl and six drops of hydrogen peroxide) was added drop by drop to the solution for 30 minutes. The stirring was continued for further 15 minutes with same condition. The dark green colour observed here confirms the formation of polyaniline. For ZNP synthesis, 0.01 g of pure ZnO nanotubes was added to reaction mixture during the aforementioned polyaniline synthesis. After 2 h of continuous stirring, the dark green precipitate was collected by centrifugation, then washed and dried for further analysis.

Synthesis of pure zinc oxide nanotubes in the presence of beta-amino ethylamine was done as already reported [17]. For the preparation of PNZ nanocomposite, 0.1 wt% commercially available emeraldine salt of polyaniline in dimethyl sulphoxide was added to the reaction mixture during the formation of ZnO nanotubes. Six hours of continuous heating at 60°C and stirring at 900 rpm resulted in a dark blue hybrid precipitate of PNZ. The precipitate was collected by centrifuging at 12000 rpm. The washed sample was then air dried and used for further characterisations.

5.2.2 Photocatalysis

To compare the activity of photocatalysts, 0.001 wt% of aqueous methylene blue (MB) dye solution was taken as the model system, and the degradation of the dye was monitored within a time frame of 70 minutes; 0.04 g of photocatalyst was added into 50 mL of the dye solution. This suspension was kept under dark for 30 minutes with stirring to avoid any error due to initial adsorption effect. The stirring in darkness also helps dye adsorption on the surface of the photocatalyst to establish adsorption–desorption equilibrium. After dark analysis, the dye solution was placed inside a photocatalytic reactor under 300W Xenon lamp. At 15 minutes of periodic exposure, 5mL aliquots were taken to monitor the dye degradation. The UV-visible absorption spectra were recorded after centrifuging at 10000 rpm for 10 minutes to avoid any scattering effect. The adsorption performance of SP in dark condition was checked by dispersing 0.05g powder in 50 mL 20 ppm MB solution.

5.2.3 Quantification of Hydroxyl Radicals from Nanocomposites

The reactive hydroxyl radicals generated from PNZ and ZNP were estimated and compared by fluorescence measurements. For the analysis, 0.005 g of PNZ or ZNP was dispersed in 50 mL alkaline 5×10^{-4} M, terephthalic acid (TTA) solution. This solution was placed under light exposure for 10 minutes with continuous stirring followed by centrifugation to get the supernatant. Fluorescence spectra of the supernatants were taken by exciting at 315 nm and compared the emission intensity.

5.3 Results and Discussions

The schematic illustration of the synthesis of PNZ and ZNP are depicted in Figure 5.1, and the mechanism of the formation of polyaniline is illustrated in Figure 5.2 [1]. The synergetic effect of Fe^{3+} ions and OH° free radicals from $FeCl_3/H_2O_2$ oxidising system oxidises aniline [A] to radical cations [B] [19]. This radical cation exists in three resonance structures as shown in Figure 5.1 [2]. These N- and para-radical cations undergo coupling to form dicationic dimer [C] [20]. The dicationic dimer undergoes rearomatisation [D] which leads to chain propagation in the presence of excess radical cations [E] and finally results in the formation of green-coloured emeraldine salt of polyaniline [F].

During the synthesis of ZNP, direct oxidation and in situ polymerisation of aniline in the presence of Fe^{3+}/H_2O_2 occurs initially and the addition of ZnO leads to the final product (ZNP). When ZnO nanotubes were added during the synthesis, a strong interaction develops between the particles and polymer due to the presence of polarons present in the polyaniline. PNZ was

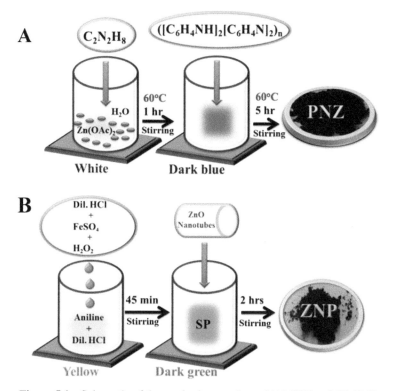

Figure 5.1 Schematic of the synthesis procedure of (A) PNZ and (B) ZNP.

synthesised by the addition of polyaniline during ZnO formation. The mechanism involved in the formation of pure ZnO nanotube was reported elsewhere [17]. Polyaniline is completely soluble in DMSO through hydrogen bond formation [18], and this polyaniline–DMSO solution when added to the alkaline reaction mixture, deprotonation of emeraldine salt (ES) to emeraldine base (EB) occurs. With the assistance of mild temperature, polyaniline in its emeraldine base structure interaction with ZnO leads to the formation of hybrid nanocomposite PNZ.

The crystallinity and phase purity of the nanocomposites are analysed from XRD pattern (Figure 5.3). Commercially available green coloured emeraldine salt of polyaniline (ES) shows amorphous nature, while the synthesised polyaniline (SP) is highly crystalline. A small percentage of ZnO powder in ZNP makes markable difference in the intensity of SP peaks without showing any peak due to ZnO, whereas in PNZ the crystalline

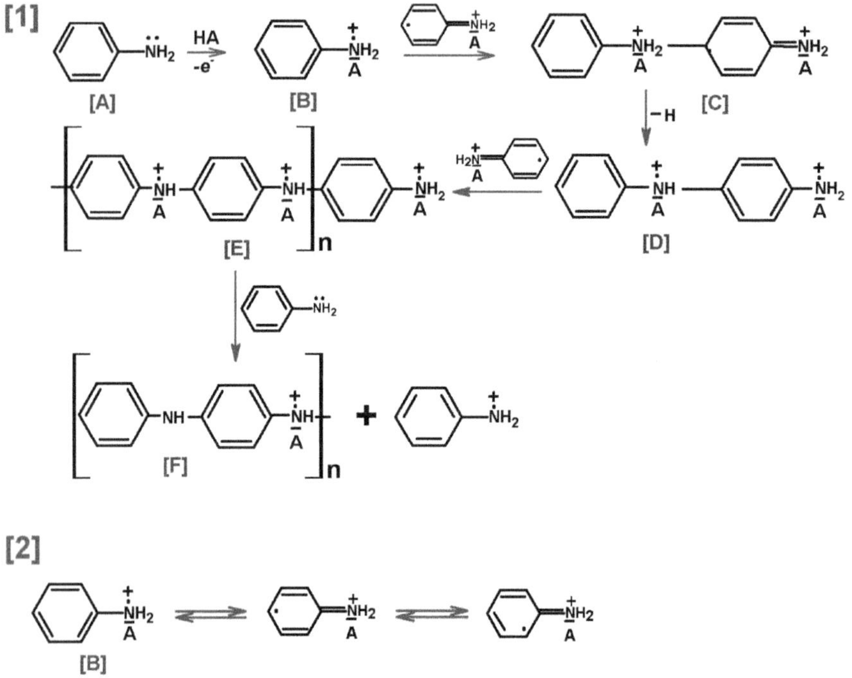

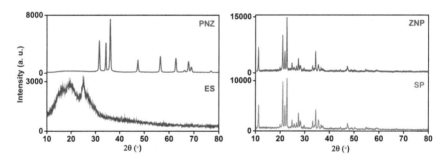

Figure 5.2   Proposed mechanism of polyaniline formation.

Figure 5.3   XRD pattern of ES, PNZ, SP and ZNP.

peaks observed are due to pure ZnO (JCPDS - 89-0510) only. This indicates that the wurtzite crystal structure of ZnO is well maintained in the PNZ composite [15].

SEM images of PNZ and ZNP clearly show the complete coverage of ZnO nanotube surface with polyaniline (Figure 5.4a and b). ES (Figure 5.4c) appears as extended fibres, whereas SP exhibited fused particle

Figure 5.4 SEM images of (a) PNZ, (b) ZNP, (c) ES and (d) SP.

morphology as seen in Figure 5.4d. The images clearly display the morphological difference in the commercially available and laboratory synthesised polyaniline.

TEM images of PNZ and ZNP are shown in Figure 5.5. Zinc oxide has been grown as short nanotubes and the attachment of polyaniline over ZnO nanotube surface are legible from the images. The existence of interface in both nanocomposites is visible from TEM images. Because of the polyaniline attachment on the surface of ZnO nanotubes, the lattice fringes of ZnO are not clearly observed.

FTIR spectra (Figure 5.6) provide ample evidence about the formation of hybrids compared to their polymer matrix. The successful polymerisation of aniline is also observed from FTIR. Intermolecular (3450 cm^{-1} / 3406 cm^{-1}) and intramolecular (2933 cm^{-1}) hydrogen bondings are present in both ES and PNZ [21]. The formed PNZ attains an emeraldine base structure in the presence of alkali during synthesis [22]. The structure of emeraldine salts have only benzenoid rings, whereas that of emeraldine base features both benzenoid and quinoid rings. C=C stretching band in both quinoid

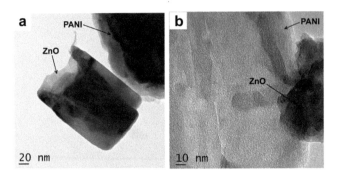

Figure 5.5   TEM images of (a) PNZ and (b) ZNP.

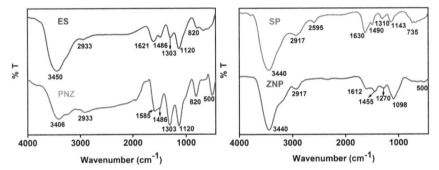

Figure 5.6   FTIR spectra of ES, PNZ, ZNP and SP.

(1585 cm^{-1}) [23] and benzenoid (1486 cm^{-1}) [24] is present in PNZ. Zn-O stretching (500 cm^{-1}) is also seen with PNZ [25]. As vibrational frequency increases in accordance with the bond order, C-N stretching frequency in quinoid (1303 cm^{-1}) [26] is found to be higher in PNZ and that attributes the formation of EB. The peaks at 1120 cm^{-1} and 820 cm^{-1} in PNZ attribute aromatic C-H group in plane and the out of plane deformation for the 1,4-disubsituted benzene. These two peaks are more intense in PNZ than ES due to the change of chemical environment [27]. These observations indicate that the ZnO crystals hold strong coulombic interaction with positively charged backbone of the polymer [28]. Besides, hydrogen bonding between imine group of polyaniline and surface hydroxyl groups of ZnO nanomaterials is also observed [29]. Both SP and ZNP have two peaks commonly at 3440 cm^{-1} and 2917 cm^{-1} due to the presence of intermolecular hydrogen bonding and N-H stretching, respectively. Protonated N-H stretching band observed at 2595 cm^{-1} in SP confirms the emeraldine salt form [30]. Peaks

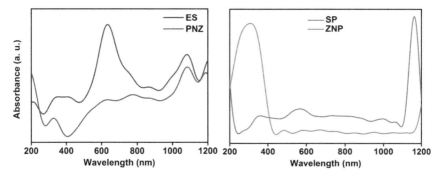

Figure 5.7 Absorbance spectra of ES, PNZ, ZNP and SP.

at 1630 cm^{-1}, 1490 cm^{-1}, 1310 cm^{-1} and 1143 cm^{-1} depict C=C stretching in disubstituted alkene, C=C bond vibration of quinoid ring, C-N stretching in amine and vibrational mode of polyaniline in emeraldine salt form, respectively [31, 32]. Shifting of these peaks to lower wavenumbers attributes the existence of strong chemical interaction between ZnO and polyaniline polymer backbone [33].

The UV-Vis-NIR spectra of synthesised hybrid samples and corresponding polymer matrices are shown in the Figure 5.7. Both ES and PNZ show a peak at 330 nm attributed to the Π-Π* transition of benzenoid ring [34], and in ES, this band appears wider between 280 to 460 nm as it contains only benzenoid rings, whereas in PNZ, the broad band at 280-400 nm attributes the incorporation of ZnO. The absorption spectrum of ES shows high intensity peak at 635 nm, due to the delocalisation of polarons. These polarons are responsible for the conduction of ES. But in PNZ, this absorption peak is fully diminished indicating the conversion of ES to EB. During the formation of EB, ES combines with ZnO and beta-amino ethylamine, resulting in the destruction of amine-associated polarons. The near IR peak at 1082 nm present in ES and PNZ attributes to the existence of polarons in amine and imine nitrogen, respectively [35]. The Π-Π* transition in SP appears broad from 310 to 390 nm absorption [34]. Other characteristic absorptions observed at 480-620 nm and 980 nm depict polyaniline in its doped form (emeraldine salt) [36, 37]. The presence of band edge at 336 nm explicitly reveals the presence of ZnO in ZNP. Characteristic absorptions (447 nm and 1082 nm) corresponding to emeraldine salt are also observed with ZNP [38].

The emission behaviour of the samples is examined from photoluminescence measurements (Figure 5.8) with an excitation wavelength of 320 nm.

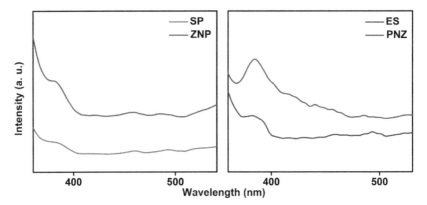

Figure 5.8 Photoluminescence spectra of SP, ZNP, ES and PNZ.

Incorporation of ZnO intensifies the emission of PNZ and ZNP compared to ES and SP which is attributed to the extended conjugation present along benzenoid and quinoid rings [26], which helps delocalisation of excitons. According to Pandiyarajan et al., the enhancement of photoluminescence in PANI/ZnO nanocomposites attributes transfer of photogenerated charge carriers from ZnO to polyaniline [39].

The size distribution studies of the samples were analysed using dynamic light scattering (DLS) experiment. The prepared samples were dispersed in methanol and subjected to laser illumination. The intensity of the scattered light from the suspension has been analysed as a function of time. Graph from Figure 5.9 shows the distribution of hydrodynamic size of the synthesised samples. PNZ exhibit reduced size compared to ZNP which again reinforces the inference from SEM image.

The adsorptivity of SP and photocatalytic activity of the prepared nanocomposites was examined with MB as a model dye. Almost 10% 20 ppm methylene blue was adsorbed on 0.05 g SP within 1 h stirring under dark. The absorption result of nanocomposites strongly recommends the visible light photocatalytic activity of polyaniline–ZnO nanocomposites [40]. The surface adsorption capability of the nanocomposites was determined by stirring with the dye solution in dark for 30 minutes. This dark adsorption analysis can eradicate the initial adsorption error and helps to establish an adsorption–desorption equilibrium. MB dye undergoes photolysis slightly in the absence of photocatalyst [41]. In the presence of photocatalyst, after exposure to light radiation, the absorption peaks of MB were gradually decreased with time. This degradation is executed by the type-II heterojunction present between

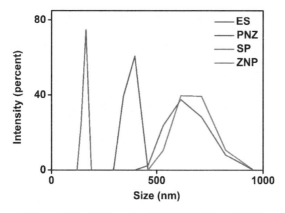

Figure 5.9   DLS graph of ES, PNZ, SP and ZNP.

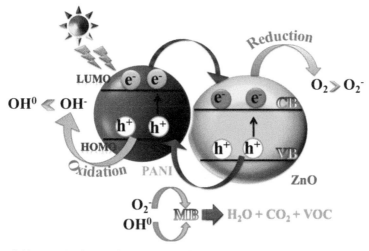

Figure 5.10 Mechanism of photocatalysis in the presence of polyaniline/ZnO nanocomposites.

polyaniline and ZnO [33]. The schematic of the possible mechanism of MB degradation in the presence of ZnO–polyaniline nanocomposites is presented in Figure 5.10. Highest MB degradation using 0.1 g ZnO/polyaniline nanocomposite is reported within 80 minutes [33]. In the present context, within 70 minutes degradation happens with 0.04 g nanocomposite. The photocatalytic degradation graph is depicted in Figure 5.11.

The presence of type II staggered heterojunction formed between polyaniline and ZnO plays a significant role in photocatalysis. When the surface

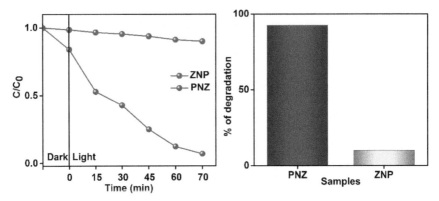

Figure 5.11   Photocatalytic degradation of methylene blue in the presence of PNZ and ZNP.

of nanocomposite was exposed to photons, photoelectrons and holes were produced in polyaniline. The excited photoelectrons transfer from HOMO to LUMO of polyaniline. Simultaneously, an interjunction transition of photo-electrons and holes occurs between polyaniline and ZnO. The photo-excited electron from polyaniline goes to conduction band (CB) of ZnO, and at CB, these electrons help for the formation of superoxides by adsorbing on O_2 molecules. At the same time, hydroxyl free radicals are generated at the valance band site. These superoxides and hydroxyl free radicals act as potential oxidising agents for the degradation of MB dye [42]. When comparing the photocatalytic performance of nanocomposites, PNZ gives better degradation compared to ZNP. PNZ appears ~12 times more photocatalytically efficient than ZNP. This may attribute to the increased adsorption of dye molecules on relatively smaller sized PNZ. Additionally ZNP has higher polyaniline content than PNZ. Methylene blue being a non-sulphonated dye adsorption over the polyaniline surface is difficult leading to reduced dye adsorption [43]. The majority of MB adsorption on polyaniline surface happens through imine nitrogen which is favourable for the increased adsorption on PNZ surface than ZNP [44]. Broad visible light absorption of PNZ also increases the photocatalytic activity. Actually, photocatalysis mainly depends on the reactive oxygen species generation from corresponding photocatalysts. Reactive oxygen species generated from PNZ and ZNP are compared by estimating the formation of 2-hydroxy terephthalic acid (HTTA) from terephthalic acid (TTA). Non fluorescent terephthalic acid is converted in to fluorescent 2-hydroxy terephthalic acid in the presence of hydroxyl free radicals (OH^o). The emission intensity of HTTA in the presence of PNZ is found to be higher

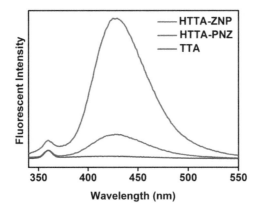

Figure 5.12   Fluorescence spectra of TTA and HTTA in the presence of ZNP and PNZ.

than that with ZNP as presented in Figure 5.12. The ratio of emission intensity from ZNP and PNZ was estimated to be 1:6, indicating high ROS generation in PNZ than ZNP. This may be attributed to the variation in composition of PNZ and ZNP and increased content of ZnO in PNZ.

5.4 Conclusions

Two types of polyaniline–ZnO hybrid nanocomposites have been success-fully fabricated by adopting the one-pot synthesis strategy. Characterisations explicitly reveal the strong interaction that exists between ZnO and polyani-line. Formation of interface between ZnO and polyaniline and the wide spectral absorption of formed nanocomposites point out the possibility for efficient photocatalytic activity. The nanocomposite synthesised with higher content of ZnO gives better degradation of methylene blue dye. Simple and versatile synthesis strategy was adopted to obtain highly crystalline polyani-line which can also be envisaged for various other applications in the near future.

Acknowledgements

Bhabhina and Jyothilakshmi are thankful to the Council of Scientific and Industrial Research (CSIR), Government of India, for the Research Fellowships. Dharsana is grateful to e-grants for the financial assistance. Sindhu thanks Kerala State Council for Science Technology and Environment (KSCSTE) for the financial support via research project. Sophisticated Test and Instrumentation Centre in CUSAT and PSGTECHS COE INDUTECH in Coimbatore are acknowledged for the TEM and SEM analysis.

References

[1] A. Pant, R. Tanwar, B. Kaur, U.K. Mandal, *Sci. Rep.,* **2018**, *8*, 14700.

[2] Y. Kim, Z. Lin, I. Jeon, T.V. Voorhis, T.M. Swager, *J. Am. Chem. Soc.,* **2018**, *140(43)*, 14413-14420.

[3] A. Malafronte, F. Auriemma, R. Di Girolamo, C. Sasso, C. Diletto, A.E. Di Mauro, E. Fanizza, P. Morvillo, A.M. Rodriguez, A.B. Munoz-Garcia, M. Pavone, C. De Rosa, *J. Phys. Chem. C* **2017**, *121*, 16617-16628.

[4] Y. Tu, N. Ahmad, J. Briscoe, D-W. Zhang, S. Krause, *Anal. Chem.* **2018**, *90,* 8708-8715.

[5] J. H. Kwon, Y. Jeon, K. C. Choi, *ACS Appl. Mater. Interfaces* **2018**, *10, 38*, 32387-32396.

[6] V. Saadattalab, A. Shakeri, H. Gholami, *Prog. Nat. Sci: Mater. Int.* **2017**, *26*, 517–522.

[7] Y. N. Jo, S. Hyun, K. Prasanna, S. Wook, C. Woo, *Appl. Surf. Sci.* **2017**, *422*, 406–412.

[8] B. N. Meethal, R. Ramanarayanan, S. Sindhu, *Appl. Nanosci.* **2018**, *8*, 1545.

[9] R. Kandulna, R. B. Choudhary, *Optik* **2017**, *144*, 40–48.

[10] S. B. Dkhil, M. Gaceur, A. K. Diallo, Y. Didane, X. Liu, M. Fahlman, O. Margeat, J. Ackermann, C. V. Ackermann, *ACS Appl. Mater. Interfaces* **2017**, *9, 20*, 17256-17264.

[11] N.-R. Kang, Y.-C. Kim, H. Jeon, S.K. Kim, J. Jang, H.N. Han, J.-Y. Kim, *Sci. Rep,* **2017**, *7*, 4327.

[12] M. Rabia, H. S. H. Mohamed, M. Shaban, S. Taha, *Sci. Rep,* **2018**, *8*, 1107.

[13] S.J. Varma, F. Xavier, S. Varghese, S. Jayalekshmi, *Polym. Int.* **2012**, 61, 743-748.

[14] B.N. Meethal, N. Pullanjiyot, Niveditha C.V., R. Ramanarayanan, S. Swaminathan, *Mater. Today. Proc.* **2018**, *5*, 16394–16401.

[15] A. Olad, R. Nosrati, *Res Chem Intermed* **2012,** 323–336.

[16] S. Chaturvedi, R. Das, P. Poddar, S. Kulkarni, *RSC Adv.* **2015**, *5*, 23563–23568.

[17] B. N. Meethal, N. Pullanjiyot, S. Swaminathan, *Mater. Des.* **2017**, *130*, 426–432.

[18] M. Bláha, J. Zedník, J. Vohlídal, *Polym. Int.* **2015**, *64*, 1801–1807.

[19] K. M. Molapo, P. M. Ndangili, R. F. Ajayi, G. Mbambisa, S.M. Mailu, N. Njomo, M. Masikini, P. Baker, E. I. Iwuoha, *Int. J. Electrochem. Sci.* **2012**, *7*, 11859–11875.

[20] S. J. Su, N. Kuramoto, *Macromolecules* **2001**, *34*, 7249–7256.

[21] N. Parveen, M.O. Ansari, M.H. Cho, *Ind. Eng. Chem. Res.* **2016**, *55*, 116–124.

[22] T. Ogoshi, Y. Hasegawa, T. Aoki, Y. Ishimori, S. Inagi, T. Yamagishi, *Macromolecules* **2011**, *44*, 7639–7644.

[23] N-A. Rangel-Vazquez, C. Sánchez-lópez, F.R. Felix, *Polímeros*, **2014**, *24*, 453–463.

[24] L. Wang, Y. Ye, X. Lu, Z. Wen, Z. Li, H. Hou, Y. Song, *Sci. rep.* **2013**, *3*, 3568.

[25] C. Pholnak, C. Sirisathitkul, S. Suwanboon, D.J. Harding, *Mater. Res.* **2014**, *17*, 405–411.

[26] V.J. Babu, S. Vempati, S. Ramakrishna, *Mater Sci Appl* **2013**, *4*, 1–10.

[27] Z. Tang, J. Wu, M. Zheng, Q. Tang, Q. Liu, J. Lin, J. Wang, *RSC Adv.* **2012**, *2*, 4062-4064.

[28] P. Rajakani, C. Vedhi, Int. *J. Ind. Chem.* **2015**, *6*, 247–259.

[29] S. Ameen, M.S. Akhtar, Y. S. Kim, O. B. Yang, H. S. Shin, *Colloid Polym. Sci.* **2011**, *289*, 415–421.

[30] A. Muslim, D. Malik, A. Abudu, *Polymer Science, Ser. B* **2012**, *54*, 518–524.

[31] A. Abdolahi, E. Hamzah, Z. Ibrahim, S. Hashim, *Materials* **2012**, *5*, 1487-1494.

[32] F. Meng, X. Yan, Y. Zhu, P. Si, *Nanoscale Res. Lett.* **2013**, *8*, 179.

[33] S. Sharma, S. Singh, N. Khare, *Int. J. Hydrogen Energy*, **2016**, *41*, 21088-21098.

[34] J. Gu, S. Kan, Q. Shen, J. Kan, *Int. J. Electrochem. Sci.* **2014**, *9*, 6858–6869.

[35] J. Laska, *Mater. Sci. Eng. B* **1999**, *68*, 76–79.

[36] W. Liu, J. Kumar, S. Tripathy, L. A. Samuelson, *Langmuir* **2002**, *18*, 9696–9704.

[37] N. R. Chiou, L. J. Lee, A. J. Epstein, *Chem. Mater.* **2007**, *19*, 3589–3591.

[38] S.A. Kumar, H. Bhandari, C. Sharma, F. Khatoon, S.K. Dhawan, *Polym Int* **2013**, *62*, 1192–1201.

[39] T. Pandiyarajan, R. V. Mangalaraja, B. Karthikeyan, *Spectrochim. Acta - Part A Mol. Biomol. Spectrosc.* **2015**, *147*, 280–285.

[40] R. Saravanan, E. Sacari, F. Gracia, M. Mansoob, E. Mosquera, V. Kumar, *J Mol Liq* **2016**, *221*, 1029–1033.

[41] P. Y. Kuang, Y. Z. Su, K. Xiao, Z.Q. Liu, N. Li, H. J. Wang, J. Zhang, *ACS Appl. Mater. Interfaces*. **2015**, *7*, 16387–16394.

[42] S. Bose, D. Dey, S. Banerjee, G. Ahmad, S. Mandal, A. K. Barua, N. Mukherjee, *J. Mater. Sci.* **2017**, *52*, 12818–12825.

[43] D. Mahanta, G. Madras, S. Radhakrishnan, S. Patil, *J. Phys. Chem. B*. **2008**, *112*, 10153–10157.

[44] M.M. Ayad, A.A. El-nasr, *J. Phys. Chem. C* **2010**, *114*, 14377–14383.

Part III

Bio-polymers

6

Synthesis and Characterisation of Polyurethanes from Bio-Based Vegetable Oil

D. Venkatesh and V. Jaisankar*

PG & Research Department of Chemistry, Presidency College (Autonomous), Chennai, Tamil Nadu, India
E-mail: vjaisankar@gmail.com
*Corresponding Author

Bio-renewable polyurethanes (PUs) are one of the most versatile class of materials today and are in demand as a high-performance industrial material due to universal availability and low cost. Vegetable oils and their derivatives have been widely used for the preparation of various polymers including polyols and polyurethanes. In this investigation, certain bio-based polyol and polyurethanes are prepared using sesame oil through epoxidation and ring opened by castor oil fatty acid. The polyurethanes are prepared by polyols (SCOL), and the isocyanates are prepared by polycondensation reaction. The prepared cross-linked polyol and polyurethanes are characterised by Fourier transform infrared spectroscopy (FT-IR) and nuclear magnetic resonance spectroscopy (^{1}H NMR), and the effect of cross-liking density and stability of the prepared polyols and polyurethanes (PUs) are investigated by thermal analysis.

6.1 Introduction

Polyurethanes (PUs) have been used in a wide range of properties which make them an indispensable material in construction, coating, adhesives, sealant, foams, elastomers, and others. Polyurethane is known for its recurring urethane linkage in the main chain and is commonly synthesised by mixing two components: one hydroxyl bearing monomer and a monomer with isocyanate groups. Considering the fact that polyol contributes the majority of the weight of polyurethane and makes full or partial substitution of polyol from bio-renewable resource in order to increase the bio-content of PUs is a promising approach [1-4]. As vegetable oils are one of the most readily available alternative renewable resource and the functional groups present in natural oils can be activated for condensation polymerisation, they converted natural oils into the polymer chain. Triglycerides of vegetable oils with the structure of an ester are formed from glycerol and three fatty acid; there are six common fatty acids of which two are saturated one, palmitic (C 16:0) and steric (C 18:0), and four are unsaturated ones, oleic (C 18:1), linoleic (C 18:2), linolenic (C 18:3) and recinolic (C 18:1 OH). Reactive sites are ester bond, carbon–carbon double bonds, and hydroxyl group which exist in some oils [4–8]. Vegetable oil polyols can be obtained by the introduction of hydroxyl group at the position of double bond by various methods. Some of the methods are hydroformulation, epoxidation followed by epoxide ring opening, hydrolysis, ozonolysis and hydrogenation. Epoxidation of vegetable oils, followed by oxirane ring opening, is one of the most important reactions to provide various polyols [8–10].

In this investigation, castor oil-based fatty acid (ICFA) was used to facilitate ring opening of epoxidised sesame oil (IESO) using a green, solvent-free/catalyst-free pathway. The polyol obtained had a higher cross-linking density than the polyol prepared from epoxy ring opening using small molecule, resulting in better thermal properties of the final polyurethanes. The effect of the molar ratio of the carboxyl to epoxy groups, reaction time and reaction temperature on the final polyol and polyurethanes (PUs) structure and functionality was investigated.

6.2 Materials and Method

Sesame oil was purchased from local source, magnesium sulphate ($MgSO_4$), hydrogen peroxide, methyl ethyl ketone (MEK) and ethyl ether were purchased from Sigma-Aldrich (Milwaukee, WI). Castor oil, hydrochloric acid, sodium hydroxide, sodium bicarbonate, formic acid, hexamethylene diisocyanate (HMDI) and 4,4 methylene diphenyl diisocyanate (MDI), and dibutyltin dilaurate (DBTDL) were obtained from Sigma-Aldrich (Milwaukee, WI). All materials were used as received without further purification.

6.2.1 Synthesis of the Epoxidised Vegetable Oil

Epoxidised vegetable oils with different epoxy groups were prepared according to a solvent free solvent method, earlier reported in Venkatesh et al [4]. Briefly, sesame oil and formic acid were charged into a 500 mL flask at 50°C to 60°C under vigorous stirring. Then, hydrogen peroxide was added slowly using syringe over a period of 5 h [4]. The reaction was continued at 50°C for another 3 h. Then, sodium bicarbonate was added to neutralise the solution, and diethyl ether was added, resulting in two layers. The organic layer was washed with distilled water until the solution become neutral, after drying with $MgSO_4$, and filtered. The organic solvent was removed by rotary evaporation and dried in a vacuum oven overnight.

6.2.2 Synthesis of Castor Oil Fatty Acid

Castor oil was saponified into fatty acid with sodium hydroxide solution at 70°C, followed by neutralisation with acid catalyst. Finally, castor oil fatty acids (ICFA) were obtained after the organic layer was purified by the same process.

6.2.3 Synthesis of Vegetable Oil-Based Polyol

The polyols were prepared by ring-opening reactions between epoxidised vegetable oils and ICFA. The polyols were identified as sesame–castor oil (SCOL). Initially, ICFA and epoxidised vegetable oil were mixed in a flask with a magnetic stirrer and maintained at 180°C at dry nitrogen atmosphere. After 7 h, a dark viscous liquid was obtained.

6.2.4 Preparation of Polyurethanes

The PUs were synthesised through the reaction of polyol with a 5% molar excess of hexamethylenediisocyanate (HMDI) in addition of two drops of DBTDL. The viscosity of the mixture was reduced by the use of MEK. The solution was heated to 70°C and stirred continuously for 3 h. The product polyuretane (HSCP) is characterised by analytical techniques. The same procedure was followed to prepare 4,4-biphenylmethylene diisocyanate based polyurethanes (BSCP).

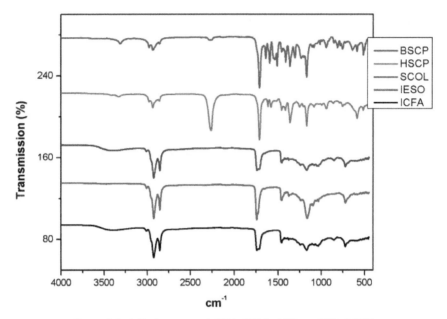

Figure 6.1   FT-IR spectra of ICFA, IESO, SCOL, HSCP, BSCP.

6.3 Characterisation

Chemical structural studies were conducted using FT-IR spectrometer with ATR platinum–diamond sampling module. A Varian spectrometer (Palo Alto, CA) at 400 MHz was used to record the ^{1}H NMR spectroscopic analyses of the monomers and final products. The thermal properties of synthesised polyol and polyurethanes were investigated by differential scanning calorimetry (DSC) (TA instrument Q200 V24.4 Build 116) and thermogravimetric analysis (TGA) (TA instrument Q500 V20.10 Build 36).

6.4 Result and Discussion

6.4.1 Infrared Spectroscopy

In this part, IR spectra of the prepared epoxidised oil, polyol (SCOL) and aliphatic and aromatic diisocyanate based polyurethanes are shown in the Figure 6.1. The prepared sesame oil based epoxidised oil, epoxy peak appeared at 826 cm^{-1}, and after the ring opening of the ICFA (fatty acid), the polyol of hydroxyl (−OH) broad peak were observed between 3000 cm^{-1} and

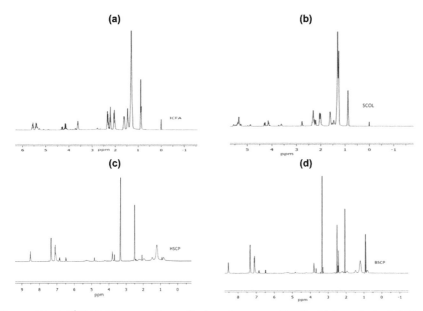

Figure 6.2 (a) ^{1}H NMR spectra showed hydroxyl content of ICFA, (b) the polyol of (SCOL), and (c) and (d) PUs of BSCP and HSCP, respectively.

3500 cm^{-1}. The functionality of prepared BSCP and HSCP were confirmed by the FT-IR spectrum. The NCO peak appeared at 2231 cm^{-1} and 2263 cm^{-1}, and the corresponding NH (stretching), C=O, and C-O vibrational peak at 3310 cm^{-1}, 1713 cm^{-1} (non-hydrogen bonded), 1641 cm^{-1} and 3329 cm^{-1}, 1714 cm^{-1} (non-hydrogen bonded), and 1629 cm^{-1} respectively, which indicate the newly synthesized product having urethane (NHCOO) group [11–14]. The observed >N-H bending vibrations at 1598 cm^{-1}, C-O-C stretching absorption band corresponding to a linkage between −OH and −NCO groups to form urethane bond in the range 1057–1130 cm^{-1} also provide strong evidence for the formation of polyurethanes. The peaks corresponding to the absorption of >NH, >C=O and >C-O- were observed at 3314, 1707 (non-hydrogen bonded), 1643 (hydrogen bonded) and 1225 cm^{-1}, respectively, which indicate the newly synthesized product having urethane (-NH-C-O-O) group [15–16].

6.4.2 Proton (^{1}H) Nuclear Magnetic Resonance Spectroscopy

Figure 6.2(a) and (b) represents the (^{1}H) NMR of ICFA and SCOL, the proton of the proton of the –CH$_2$COO– signal appears at 2.21 ppm, the methylene proton of the carbon–carbon double bond signal is observed at 3.35 ppm, the terminal and long-chain protons of CH_3 and CH_2CH signal is observed at 0.89 ppm and 1.63 ppm, respectively.

The SCOL of C=C proton intensity reduce when oxirane ring opened by castor oil based fatty acid (ICFA), the intensity of peak is 5.30 ppm. shows, the different environment of fatty acid chain in hydroxyl group, the signal appeared at 3.65 ppm, after the reduction reaction the peaks at 3.0 ppm disappeared [17–23]. On the other hand, the increased peak intensity at 3.5–3.6 ppm is due to the complete reduction of triglyceride. This leads to the formation of primary –OH groups backbone which appeared at 5.54 ppm, resulting in the formation of primary –OH groups backbone appears at 5.54 ppm. Hence, the respective peaks confirms the formation of SCOL (Figure 6.2b). The polycondensation of polyol (SCOL) reacts with two different isocyanates resulting in the formation of polyurethane was observed by the reduction in intensity of polyol peak at 5.31 ppm, the intensity of polyol peak at 5.31 ppm decreased, and the new signal of urethane (NH) was shown at HSCP and BSCP 8.53 ppm, other peak observed (from the MDI based PUs) peaks observed at 7.0–7.10 ppm, and $-CH_3$ peak individually observed at 3.80 ppm designated to the protons of CH_2OCO support the synthesis of polyurethane from the blend of polyol and isocyanate (SCOL). The other peaks observed were assigned to C=C proton as 5.26 ppm (Figure 6.2c,d) [24–25].

6.4.3 Thermal Analysis

The differential scanning calorimetry was used to study the phase transition occurring during the heating of SCOL polyol and polyurethanes (PUs). The DSC thermograms of polyol Perkin–Elmer DSC were obtained by heating sample from $-80°C$ to $200°C$ at a heating rate of $20°C$ min^{-1} under nitrogen atmosphere. This was illustrated in Figure 6.3a.

The DSC curve showed that the polyol mixture of SCOL had a melting point at $-7.58°C$ and then the glass transition temperature at $-56.66°C$. The observed melting point of SCOL has a small difference in this fatty acid chain; the –OH number of the fatty acid is different but the instance sesame–castor oil polyols is higher than the hydroxyl number. In general, the prepared polyol has the potential to produce rigid lipid-based polyurethanes. The acidity of polyol might be attributed to the formation of fatty acid during the preparation. Usually, acid number below 10 mg KOH/g was considered low enough for commercial polyols [26]. The DSC thermograms

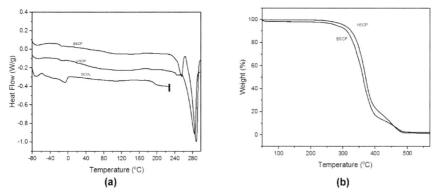

Figure 6.3 (a) and (b) show DSC and TGA curve of polyol and polyurethanes.

of the all PUs are shown in Figure 6.3a. The prepared polyurethanes of glass transition temperature (Tg) were in the range of HSCP and BSCP at $-80°C$ to $300°C$. The decomposition values are recorded and compared with HSCP and BSCP polyurethanes [27–28]. However, the PUs prepared from chain extender of HMDI and MDI melt in the range of $220°C$ and $245°C$. Glass transition temperature observed in the first and second heating of PUs samples, HSCP and BSCP, at $235°C$, $284°C$ and $255°C$, $287°C$, respectively. The second glass transition temperature (Tg) is closer to the PUs of HSCP and BSCP which indicated that the components were compatible and the PUs were amorphous. A weak melting point observed in HSCP, which implies the existence of phase separation in the sample, though the sample stands transparent. For sample BSCP, a melting point was clearly observed around $295°C$ and the sample was opaque due to phase separation. Thermal analysis of the HMDI- and MDI-based polyurethanes exhibits a temperature gap at $100°C$. The second degradation BSCP ranges approximately $470°C$ and attributes to the amide degradation. By comparing HSCP and BSCP polyurethanes, the first decomposition at $270°C$ and $260°C$. The second decomposition range observed at $410°C$ and $400°C$ imply that the samples were completely decomposed at $520°C$. The low yield was observed mostly due to the usage of HMDI which contains aliphatic chain whereas high yield was obtained for aromatic chain extender MDI based PU [28–29]. The improved stability was assumed to be due to the MDI-based urethane linkage of recinolic acid, and the observed temperature ranges might be explained by degradation of HMDI and MDI functions.

6.5 Conclusions

Vegetable oil-based polyurethanes have been prepared using polyols and polyurethanes synthesised from sesame oil with aromatic 4,4 methylene diphenyl diisocyanate and aliphatic hexamethylene diisocyanate. These polyols and polyurethanes were confirmed by FT-IR and proton NMR spectroscopy. The stability of HSCP and BSCP-based polyurethanes are characterised by TGA and DSC. The addition of chain extender MDI exhibits higher stability. The prepared bio-based polyurethanes can be used for various applications such as, biomedical, aerospace, and energy storage devices.

References

[1] Chen, R., Zhang, C. and Kessler, M. R. J. Appl. Polym. Sci., **2014**, 132, 41213.

[2] Reena Vaidya, Gajanan Chaudhary, Nitin Raut, Ganesh Shinde and Nilesh Deshmukh, IPCBEE., **2012**, 41.

[3] Jing Zhang, Ji Jun Tang, and Jiao Xia Zhang, Int. J. Polym. Sci., Volume **2015**, Article ID 529235.

[4] Venkatesh, D.; Jaisankar, V.; Inter. J. Sci. Eng. Invest. **2018,** 7(73), 44-51

[5] Mohammed A. Mekewi, Ahmed M. Ramadan, Farida M. ElDarse, Mona H. Abdel Rehim, Nabawea A. Mosa, Mahmoud A. Ibrahim, Egy. J. of Petr., **2016**,02,002

[6] Xiaohua Kong, Suresh S, Narine, ACS Biomacromolecules, **2007**, 8, 2203-2209.

[7] Mohd Zan Arniza., Seng Soi Houng., Zainab Idris., Shoot Kiam Yeong., Hazimah Abu Hassan., Ahmed Kushairi Din., Yuen May Choo., J. Am. Oil., Chem. Soc., **2015**, 92, 243-255.

[8] Chaoqun Zhang, Samy A. Madbouly, and Michael R. Kessler, ACS Appl. Mater. Interfaces, **2015**, 7, 1226-1233.

[9] Honghai Dai, Liting Yang, Bo Lin, Chengshuang Wang, Guang Shi, J. Am. Oil., Chem. Soc., **2009**, 86, 261-267.

[10] Therefrom Chaoqun Zhanga, Ying Xiaa, Ruqi Chena, Seungmoo Huhe, Patrick A. Johnstonc, Michael R. Kessler a, b, d, published on 12 April 2013.

[11] Zoran S. Petrovic, Liting Yang, Alisa Zlatanic, Wei Zhang, Iv an Javni Kansas, J. Appl., Pol, Sci., **2007**, 105, 2717–2727.

[12] Sylvain Caillol, Myriam Desroches, Gilles Boutevin, Ce dric Loubat, Re mi Auvergne, and Bernard Boutevin, Eur. J. Lipid Sci. Technol. **2012**, 114, 1447-1459.

[13] Mandar, S, Gaikwad, Vikas V, Gite, Pramod P. Mahulikar, Dilip G, Hundiwale, Omprakash S. Yemul, Prog. In Org. Coat., **2015**, 86, 164-172.

[14] Petrovic, Z. S.; Liting Yang.; Zlatanic, A.; Zhang, W.; Kansas, J.; J. Appl. Poly. Sci., **2007,** 105, 2717–2727.

[15] Caillol, S., Desroches, M., Boutevin, G., Loubat, C.D., Auvergne, R.M., Boutevin, B., Eur. J. Lipid Sci. Tech., **2012,** 114,1447-1459.

[16] Zhang, C., Xia, Y., Chen, R., Kessler, M.R., ACS Sus. Chem., **2014,** 2, 2465-2476.

[17] Meier, Metzger, M.A.R., Schubert, J. O., U.S. Chem. Soc. Rev., **2007,** 36, 1788-1802(2007).

[18] Arniza, M.Z., Hoong, S.S.; Idris, Z.; Yeong, S.K.; Hassan, H.A.; Din, A.K.; Choo, Y. M.; J. Am. Oil Chem. Soc., **2015,** 92, 243-255.

[19] Rajendran, T. V.; Jaisankar, V.; J. Mat., **2015,** 2, 4421-4428.

[20] Zhang, C., Xia, Y., Chen, R., Huh, S., Johnston, P.A., Kessler, M.R., ACS Sus. Chem., **2013,** 152, 1477-1484.

[21] Okieimen, F.E., Bakare, O.I., Okieimen, C.O., Ind. Crops Products., **2002,**15, 139–44.

[22] Lozada, Z., Suppes, G.J., Tu, Y.C., Hsieh, F.H., J. App. Poly. Sci., **2009,** 113, 2552–2560.

[23] Suman Thakur, Niranjan Karak, Prog. Org. Coat., **2013,** 76, 157-164.

[24] Xiaohua Kong, Guoguang Liu, Jonathan, M. Curtis, Inter. J. Adhe., **2011,** 31, 559 – 564.

[25] Gurnule, W.B., Thakre, M.B., J. Phar. Bio. Chem. Sci., **2014,** 5(2), 204-213.

[26] Shadpour Mallakpour, Mehdi Taghavi, Ira. Poly. J., **2009,** 18(11), 857-872.

[27] Khan, M. I., Azizli, K., Sufian, S., Man, Z., Khan, A.S., RSC Adv., **2012,** 00, 1-3.

[28] Jing Hu, Meixia Chen, Huaixiang Tian, Weijun Deng, RSC Adv., **2015,** 5, 81134- 81141.

[29] Basta, A., Mauro Missori, Girgis, A. S., Marco De Spirito, Massimiliano Papi, Houssni El-Saied, RSC Adv., **2014,** 4, 59614-59625.

7

Application of *Lepidium sativum* as an Excipient in Pharmaceuticals

S. V. Sutar[1], S. S. Shelake[2], S. V. Patil[3] and S. S. Patil[2]

[1]Department of Pharmaceutical chemistry, Ashokrao Mane College of Pharmacy, Peth-Vadgaon, Hatkanangale, Kolhapur, Maharashtra, India
[2]Department of Pharmaceutics, Ashokrao Mane College of Pharmacy, Peth-Vadgaon, Hatkanangale, Kolhapur, Maharashtra, India
[3]Department of Pharmaceutics, Shree Santkrupa College of Pharmacy, Ghogaon, Karad, Satara, Maharashtra, India

Various types of plant mucilage available like alginic acid, gelatin maize starch and potato starch have been used as a binder in pharmaceutical formulation. But still finding a novel binder is useful in the pharmaceutical industry for manufacturing tablets. *Lepidium sativum* was chosen for its binding property. Aspirin and ibuprofen tablets were prepared by wet granulation technique using *Lepidium sativum* as a tablet binder. The prepared tablets were evaluated for physiochemical characteristics, and the binding efficacy of the *Lepidium sativum* was compared with the standard binder mucilage polyvinyl pyrrolidine (PVP) at similar concentration (3% w/w), 27.16^o to 28.45^oangle of repose and 0.46–0.46% w/w friability 1.2 to 12.03 min disintegration time. Tablets at 3% w/w binder concentration showed more optimum results as tablet binder. *Lepidium sativum* was found to be useful for the preparation of uncoated tablet dosage form. *Lepidium sativum* can be an alternative binder for the pharmaceutical formulations. Abundant availability, food grade status, economic feasibility, commercial suitability and reliability make the mucilage an alternative for the existing synthetic excipients.

7.1 Introduction

Garden cress (Gc), *Lepidium sativum* is an annual, herbaceous edible plant that is botanically related to mustard and watercress. Gc plant is native to Egypt and South west Asia. It is cultivated in India, North America and parts of Europe. In some regions, Gc is known as garden pepper grass, pepper cress, pepperwort or poor man's pepper. The plant is cultivated as a culinary vegetable all over Asia. Gc seeds are well known for their various ethno-pharmacological properties. Gc seeds are used in South Asia as a traditional medicine to treat bronchitis, asthma and cough. It is considered abortifacient, diuretic, expectorant, aphrodisiac, antibacterial, gastrointestinal stimulant, gastro protective, laxative and stomadic. Gc seed is reported to exhibit antirheumatic and bronchodilatory potential. The paste of GC seeds is applied in rheumatic joints to relieve the pain and swelling. It is also useful in hiccup, dysentery, diarrhoea and skin disease caused by impurities of blood. Ethanolic extracts of Gc seed were effective in treating inflammatory bowel disease. Traditional sweets for lactating mothers are prepared from the Gc seeds. The present investigations deal with the nutritional, nutraceutical, antimicrobial properties and phytochemical constituents of Gc seeds. It also highlights the potential of Gc seeds and its extracts for various medicinal uses. Therefore in this study, it is proposed to carryout formulation and evaluation of *Lepidium sativum* as an excipient in pharmaceutical formulation.

Physicochemical Properties of mucilage

Chemical properties: Chemically, the mucilage is an acidic complex polysaccharide composed of salts of sugars other than glucose: L-arabinose, D-galactose, L-rhamnose and D-glucuronic acid combined with certain metallic cations such as sodium, potassium, calcium and magnesium. For the most part, the mucilage has highly branched structures containing different sugar units with many possible variations as regard to degree of branching, length of branches and type of linkages. Therefore, an almost infinite number of structures are possible. Forces act between molecules, between different parts of the same molecules and between polymer and solvent. These forces include hydrogen bonding, ionic charges, dipole and induced dipole inter-actions and Vander Waals forces. All these forces affect properties as gel forming tendency, viscosity and adhesiveness.

Physical properties: The physical properties of mucilage are of the first importance in determining their uses and their commercial value. Mucilage as seen or collected in the natural state are represented in a variety of shapes and forms. They are available in spheroidal tears up to 32 mm in diameter, crystals, granules, powder (by mechanical process), spray and roller dried powder. The surface of most mucilage when fresh is perfectly smooth, but this may soon become rough or covered with minute cracks or striations due to weathering. The Colour of mucilage (in the solid state) varies from almost white through various shades of yellow, amber and orange to dark brown. Colour is of great importance in the commercial valuation of mucilage [1].

Methods of incorporation of binders: Tablet cohesion is best achieved when the binding component is used in solution as an adhesive. In solution form, the binder is well distributed in the other materials of the tablet and results in better bonding with a lower concentration of binder. Moreover, since powders differ with respect to the ease with which they can be wetted, and their rate of solubilisation, it is preferable to incorporate the binding agent in solution. Some poorly compressible drugs like paracetamol, metronidazole and acetazolamide can be successfully tableted only when a liquid adhesive and wet granulation procedure is employed. The method of liquid addition can change from pouring the total amount of liquid at once, to the pumping of liquid for a specific period of time during granulation. Binder solutions are usually made up to weight rather than volume. This enables the formulator to determine the weight of the solids, which have been added to the tablet granulations in the binding solution.

Taxonomy of *Lepidium sativum*: Garden cress (Gc), *Lepidium sativum* is an annual, herbaceous edible plant that is botanically related to mustard and watercress. Gc plant is native to Egypt and South west Asia. It is cultivated in India, North America and parts of Europe. In some regions, Gc is known as garden pepper grass, pepper cress, pepperwort or poor man's pepper. The plant is cultivated as culinary vegetable all over Asia. Gc seeds are well known for their various ethno pharmacological properties. Gc seeds are used in South Asia as traditional medicine to treat bronchitis, asthma and cough. It is considered abortifacient, diuretic, expectorant, aphrodisiac, antibacterial, gastrointestinal stimulant, gastro protective, laxative and stomadic. Gc seed is reported to exhibit antirheumatic and bronchodilatory potential. The paste of GC seeds is applied in rheumatic joints to relieve the pain and swelling. It is also useful in hiccup, dysentery, diarrhoea and skin disease caused by impurities of blood. Ethanolic extracts of Gc seed were effective in treating inflammatory bowel disease. Traditional sweets for lactating mothers are prepared from the Gc seeds. The present review deals with the nutritional, nutraceutical, antimicrobial properties and phytochemical constituents of Gc seeds. It also highlights the potential of Gc seeds and its extracts for various medicinal uses.

Following are the species:

1. *Lepidium draba*
2. *Lepidium crassifolium*
3. *Lepidium latifallum*
4. *Lepidium ruderale*
5. *Lepidium perfoliatum*

Botanical description: A variety of parts of the plant namely; seeds, leaves and roots have been used in treating various human ailments. Seeds powder were creamish yellow in colour, microscopy of the seeds powder shows uniform thick walls, oily endosperm, number of reddish-brown fragments of seed coats and reddish colouring matter.

Botanical classification: Kingdom: Planate, Division: Magnoliophyta, Class: Magnoliopsida, Order: Brassicales, Family: Brassicaceae, Genus: *Lepidium sativum*

Seeds:- The acute toxicity tests showed that the administration of extract in single doses of 0.5 to 3.0g/kg of body weight of mice did not produce any adverse effects or mortality in mice. Chronic toxicity (100 mg/kg/day for a period of 3 months in drinking water) studies revealed that there were no symptoms of toxicity except a statistically insignificant higher mortality rate in the mice [2].

7.2　Material and Methods

7.2.1　Materials

Isolation and purification of mucilage: In this method, the seeds (100 g) were soaked for 12 hour in distilled water (1 L). Then mucilage was separated by passing through vacuum pump. After that remaining particulate matter separated by passing through muslin cloth. Then separated clear material was treated with ethanol. So as to get precipitated mucilage. Drying was done at 45°C for 6 h. Then powder was passed through # 80 mesh sieve and weighed to calculate the yield.

7.2.2　Methods of Formulation

Preparation of tablets by wet granulation method: The formulation was developed by using aspirin IP as model drug. The binder concentrations used

Figure 7.1 *Lepidium sativum* plant.

Figure 7.2 Seeds of *Lepidium sativum*.

were 1, 2, 3, 4, 5 and 6% w/w. Binder level was adjusted by lowering the level of microcrystalline cellulose in the formula. All ingredients were dry mixed manually in mortar and granulated with water as granulating fluid. The wet mass was passed through sieve # 12 mesh. The granules were dried at 40°C for 1 hour in tray dryer. The dried granules were passed through sieve # 40, lubricated with lubricant sand were compressed by using hydraulic press with flat faced punches. The tablet formulation was developed for aspirin 200 and ibuprofen. Similar procedure was employed for preparation of aspirin and ibuprofen tablet using polyvinyl pyrrolidine [3].

Table 7.1 Materials used their category and pharmacopoeial status.

Name of material	Category	Pharmacopoeial status
Aspirin	Beta blocker	IP
Ibuprofen	Anti-epileptic	USP
Microcrystalline cellulose	Disintegrating agent	PH. Eur.
Lepidium sativum	Binder	PH. Eur.
Polyvinyl pyrrolidine	Binder	PH. Eur.
Magnesium stearate	Lubricant	PH. Eur.
Talc	Diluent	PH. Eur.

All drugs and other materials or chemicals used were of analytical grade.

Table 7.2 Formulation of aspirin tablets with *Lepidium sativum* as binder.

Formulation	Ingredients (mg/Tablets)					
	Aspirin (mg)	Binder *Lepidium sativum* (mg)	Microcrystalline Cellulose (mg)	Magnesium stearate (mg)	Talc (mg)	Total (mg)
A1	300	5	180	10	5	500
A2	300	10	175	10	5	500
A3	300	15	170	10	5	500
A4	300	20	165	10	5	500
A5	300	25	160	10	5	500
A6	300	30	155	10	5	500
A3(PVP)	300	15	170	10	5	500

Table 7.3 Formulation of ibuprofen tablets with *Lepidium sativum* as binder.

Formulation	Ingredients (mg/tablets)					
	Ibuprofen (mg)	Binder *Lepidium sativum* (mg)	Microcrystalline cellulose (mg)	Magnesium stearate (mg)	Talc (mg)	Total (mg)
I1	400	6	182	6	6	600
I2	400	12	176	6	6	600
I3	400	18	170	6	6	600
I4	400	24	163	6	6	600
I5	400	30	158	6	6	600
I6	400	36	152	6	6	600
I3(PVP)	400	18	170	6	6	600

7.2.3 Experimental Work

Preformulation studies of *Lepidium sativum*

Preformulation studies were performed on the *Lepidium sativum*, which included extraction, purification and physiochemical characterisation of the powder.

Physicochemical characterisation of the *Lepidium sativum*

i. Solubility study

The separated mucilage was evaluated for solubility in water, acetone, chloroform and ethanol in accordance with the British pharmacopoeia specifications.

ii. Loss on drying

The method adopted was that specified in the B.P. 2004 for acacia. 1.0 g of the sample was transferred into each of several Petri dish is and then dried in an oven at 105°C until constant weight was obtained. The moisture content was then determined as the ratio of weight of moisture loss to weight of sample express ease percentage.

iii. pH determination

pH was determined by shaking 1% w/v dispersion of the sample in water for 5 min and the reading were noted by digital pH meter.

iv. Bulk and tapped density

2.0 g quantity each of the powder sample was placed in a 10 ml measuring cylinder and the volume, Vb, occupied by each of the samples without tapping was noted. After 100 taps on the table, the occupied volume Vt was read. The bulk and tap densities were calculated as the ratio of weight to volume (Vb and Vt respectively) by the following equations;

$$BD = \text{Weight of the Powder/Volume of the packing}$$
$$TBD = \text{Weight of the powder /Tapped volume of the packing}$$

v. Hausner's Ratio

This was calculated as the ratio of tapped density to bulk density of the samples.

$$\text{Hausner's ratio} = t/d$$

When t is the tapped density and d is bulk density. Lower H(1.25)

vi. Compressibility Index (C%)

This was calculated using the equation:

Compressibility Index = Tap density – Bulk density/Tap density × 100

vii. Angle of repose

The static angle of repose "θ" was measured according to the fixed funnel and free standing cone method. A funnel was clamped with its tip 2 cm above a graph paper placed on a flat horizontal surface. The powders were carefully poured through the funnel until the apex of the cone thus formed just reached the tipoff the funnel. The mean diameters of the base of the powder cones were determined and the tangent of the angle of repose calculated using the equation [3],

$$\theta = \tan - 1(h/r)$$

Preformulation studies drugs

Preformulation studies were perform don the drug, which included melting point determination, solubility and compatibility studies.

i. Determination of melting point

Melting point of the drug was determined by taking small amount of drug in a capillary tube Closed at one end .The capillary tube was placed in a melting point apparatus and the temperature at which drug melts was recorded. This was performed thrice and average value was noted.

ii. Determination of pH

The pH of aspirin/ibuprofen was determined using potentiometer for freshly prepared 1% aqueous solution of aspirin [4].

iii. Fourier transformation infrared spectroscopy (FTIR)

The spectra were recorded for polymer, pure drug and spherical agglomerates using FT-IR. FTIR spectra were obtained using a Shimadzu FTIR spectrometer (IR Affinity 1 Model, Japan) spectrometer. The pellets of drug and KBr were prepared on KBr-press. The samples were scanned over the range of 4000–800 cm-^{1}.

iv. Compatibility

The compatibility study was done to check if drug and polymers used in formulation are compatible or not. The compatibility was checked by FTIR.

FTIR of drug and drug with polymers samples were kept in stability chamber at 400 c and 75% RH and after one month FTIR was taken.

Pre-compression evaluation

Organoleptic properties: A small quantity of the drug powder was taken in butter paper and viewed in well-illuminated place.

i. Sieve analysis: Average size of the API's was determined using vibratory sieve shaker. 50g of API was weighed and placed on an ultrasonic sieve shaker. The test was carried out at amplitude of 50 for 10 minutes. Percentage retained on #20, #40, #60, #80, #100 and fines was determined [5].

ii. Moisture content: Moisture content of the granules was determined gravimetrically by taking 5g from each batch of the granules and heating the samples in an oven at 120°C for 1 hour. The granules were weighed immediately and the loss in weight was considered as moisture content of the granules.

$$\%W = \frac{A - B}{B} \times 100$$

iii. Determination of bulk, tapped density: Both loose bulk density (ρb) and tapped bulk density (ρt) was determined. A quantity of 2 gm of granules from each formula, previously shaken of break any agglomerates formed, was introduced into 10 ml measuring cylinder. After that the initial volume was noted and the cylinder was allowed to fall under its own weight on to a hard surface from the height of 2.5 cm at 2 second intervals. Tapping was continued until no further change in volume was noted. ρu and ρt were calculated using the above formulas.

iv. Density related properties: The Carr's Index (Percent compressibility) of the granules was calculated from the difference between the tapped and bulk densities divided by the tapped density and the ratio expressed as a percentage. The Hausner ratio was calculated by dividing the tapped density by the bulk density of the granule.

v. Determination of angle of repose: The angle of repose of granules was determined by the funnel method. The accurately weight granules were taken in the funnel. The height of the funnel was adjusted in such a way the tipoff the funnel just touched the apex of the granules. The granules were allowed to flow through the funnel freely on to the surface. The diameter of the granules cone was measured and angle of repose was calculated using the above mentioned equation.

Post-compression evaluation of tablets

i. Thickness: The thickness of the tablets was determined by using Vernier calipers. Five tablets were determined, and average values were calculated.

ii. Weight variation test: The weight variation test would be a satisfactory method of determining the drug content uniformity of tablets if the tablets were all or essentially all (90 to 95%) active ingredients, or if the uniformity of drug distribution in the powder from which tablets were made perfect. Ten tablets were taken and their weight was determined individually and collectively using single pan electronic balance. The average weight of the tablets was determined from collective weight. From the individual tablets weight, the range and percentage standard deviation was calculated. Not more than 2 tablets should deviate from the average weight of tablets and the maximum Percentage of deviation allowed. In direct compression of tablet, uniform weight of tablets represents appropriate powder flow and uniform die filling [6].

iii. Hardness: Hardness indicates the ability of at able to with stand mechanical shocks while handling. The hardness of the tablets was determined using Monsanto hardness tester. It is expressed in kg/cm2.Three tablets were randomly picked and hardness of the tablets was determined.

iii. Friability test: The friability of tablets was determined using Roche Friabilator. It is expressed in percentage (%). Ten tablets were initially weighed (+F) and transferred into friabilator. The friabilator was operated at 25 rpm for 4 minutes or run up to100 revolutions. The tablets were weighed again (W). The % friability was then calculated [7].

$$\%Friability\,(F) = \frac{w_i - w_f}{w_i} \times 100$$

iv. Disintegration test: DT test was carried out according to USP specification. Tablets were placed in a disintegration tester (type USP-Electro lab USP-ED-2 AL) filled with distilled water at $37\pm0.2°C$. The tablets were considered completely disintegrated when all the particles passed through the wire mesh. Disintegration times recorded were mean of 6 determinations [8].

v. In vitro dissolution studies: The release rate of aspirin and ibuprofen from tablets was determine during The United States Pharmacopoeia (USP) XXIV dissolution test in apparatus II (paddle method). The dissolution test was performed using 900 ml of acetate buffer for aspirin and phosphate buffer for Ibuprofen, at $37\pm0.5^{\circ}C$ and 50 rpm for aspirin and 50 rpm for Ibuprofen. A sample (10 ml) of the solution was withdrawn from the dissolution apparatus, and was replaced with fresh dissolution medium. The samples diluted to a suitable concentration. Absorbance of the dissolutions was measured at aspirin 265 nm and ibuprofen 221 nm using Agilent UV-Vis double beam spectrophotometer. For aspirin and ibuprofen cumulative percentage of drug release was calculated using the equation obtained from standard curve [9].

7.3 Result and Discussion

Preformulation studies of *Lepidium sativum*

Extraction *Lepidium sativum*:

Figure 7.3 Trituration of *Lepidium sativum.*

Organoleptic properties of aspirin and ibuprofen:

Table 7.4 Organoleptic properties aspirin and ibuprofen.

Test	Observations of aspirin	Observations of ibuprofen
Colour	White crystalline powder	White crystalline powder
Taste	Bitter	Bitter
Odour	Odourless	Characteristic

Solubility: Aspirin is soluble in water and alcohol, slightly soluble in chloroform dichloromethane; practically insoluble in ether. Ibuprofen is readily soluble in most organic solvent and very soluble in alcohol.

pH: 1% w/v solution of aspirin hydrochloride in water has pH of 4.5

2% w/v solution of ibuprofen hydrochloride in water has pH of 7.2

Melting point: Melting point of aspirin was found to be 135°C and ibuprofen was found to be 75–77°C. From this we concluded that the drug sample is pure.

Sieve analysis: The data obtained from sieve analysis is tabulated in Table 7.5.

Table 7.5   Particle size distribution of the aspirin and ibuprofen.

Sieve no.	Retention % w/w of Aspirin	Retention % w/w of Ibuprofen
# 20	0.0	0.0
# 30	0.0	0.0
# 40	0.99	0.99
# 60	0.99	0.99
# 80	2.97	2.97
# 100	17.82	17.82
Through 100	18.558	18.558

FTIR spectrum

Aspirin:

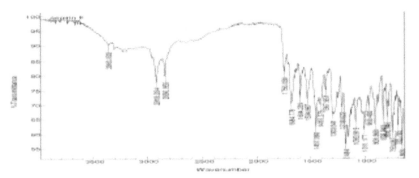

Figure 7.4   FT-IR spectrum aspirin.

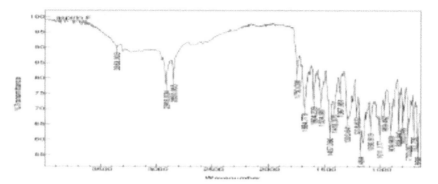

Figure 7.5 FTIR spectrum of drug and polymer mixture after one month.

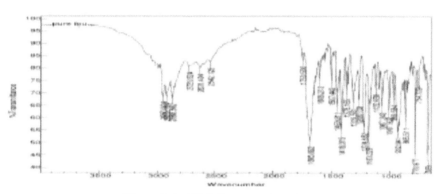

Figure 7.6 FT-IR Spectrum Ibuprofen.

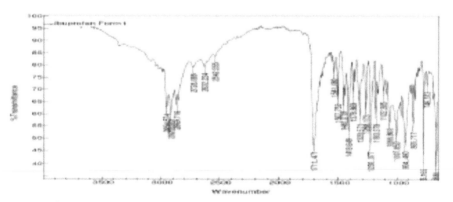

Figure 7.7 FTIR spectrum of drug and polymer mixture after one month.

The FTIR spectrum of aspirin and ibuprofen was showed in Figures 7.4 and 7.6. Major functional groups present in aspirin and ibuprofen were showed characteristic peaks in FTIR spectrum. The major peaks were identical to functional group of aspirin and ibuprofen. Hence, the sample was confirmed as aspirin and ibuprofen.

FT-IR spectra of aspirin and ibuprofen alone and its combination with polymers are shown in Figures 7.5 and 7.7. An FT-IR spectrum of pure aspirin showed the peaks 1700–1725 cm^{-1}(C=O stretching), 2500–3300 cm^{-1} (O-H stretching), 1735–1750 cm^{-1} (COO-, stretching) and 1000-1300 cm^{-1} (C-O, stretching).

An FT-IR spectrum of pure Ibuprofen showed the peaks 1700-1725 cm^{-1} (C=O stretching), 2850-3000 cm^{-1} (C-H stretching) and 2500-3300 cm^{-1} (OH, stretching). These peaks can be considered as characteristic peaks of aspirin and ibuprofen and were not affected and prominently observed in IR spectra of aspirin and ibuprofen along with polymers as shown in the Figures 7.5 and 7.7 indicated no interaction between aspirin and ibuprofen with excipients.

Evaluation

Precompression evaluation

Aspirin:

Table 7.6   Density and related properties of *Lepidium sativum* powder determined at different binder concentration.

Formula binder weight (%w/w)	Bulk density (g/ml)	Tapped density (g/ml)	Carr's Index (%)	Hausner's ratio	Angle of repose(θ)
A1	0.2301±0.25	0.2802±0.36	21.72±0.84	1.22	31.96
A2	0.2347±0.21	0.2731±0.24	16.36±0.80	1.16	30.54
A3	0.2398±0.23	0.2697±0.31	12.46±0.65	1.12	27.16
A4	0.2398±0.36	0.2631±0.28	9.71±0.87	1.09	25.84
A5	0.2425±0.18	0.2628±0.30	8.37±0.56	1.08	25.32
A6	0.2429±0.24	0.2600±0.32	8.2±0.76	1.07	24.50

The bulk density was found to be 0.2201g/ml to 0.2329 g/ml and tapped density 0.2701 to 0.2500. The results of Carr's Consolidation Index or Compressibility Index (%) for the entire formulation are from 17.83 to 7.0. They had shown excellent compressibility index values up to 9.12 % result in good to excellent flow properties. The data obtained for angle of repose for all the formulations were tabulated in Table 7.6. The values were found to be in the range of 30.96 and 23.50(θ) They had shown excellent compressibility index values up to 29.54 (θ).

Ibuprofen:

Table 7.7 Density and related properties of powder determined at different binder concentration.

Formula binder weight (%w/w)	Bulk density (g/ml)	Tapped density(g/ml)	Carr's Index (%)	Hausner's ratio	Angle of repose(θ)
I1	0.2601±0.21	0.3102±0.24	19.26±0.45	1.19	33.02
I2	0.2642±0.25	0.3041±0.32	15.10±0.57	1.15	32.37
I3	0.2664±0.20	0.2947±0.34	10.62±0.54	1.10	28.45
I4	0.2686±0.26	0.2930±0.36	9.08±0.50	1.09	25.52
I5	0.2686±0.28	0.2903±0.25	8.07±0.47	1.08	23.26
I6	0.2731±0.27	0.2877±0.26	5.34±0.41	1.05	22.04

The bulk density was found to be 0.2664g/ml to 0.2631 g/ml and tapped density 0.32697 to 0.2977. The results of Carr's Consolidation Index or Compressibility Index (%) for the entire formulation are from 10.62 to 7.1. They had shown excellent compressibility index values up to 9.64 % result in good to excellent flow properties. The values of angle of repose were found to be in the range of 28.45 and 23.50(θ). They had shown excellent compressibility index values up to 27.45 (θ).

Post-compression evaluation

Aspirin:

Table 7.8 Weight variation, thickness and hardness of aspirin.

Formulation code	Weight variation (%)	Tablet thickness (mm)	Hardness kg/cm2
A1	0.47%	1.6	3.9±0.38
A2	0.58%	1.5	5.6±0.34
A3	0.34%	1.6	6.2±0.32
A4	0.65%	1.5	7.6±0.28
A5	0.66%	1.5	8.8±0.26
A6	0.77%	1.6	9.1±0.24

All the tablets passed weight variation test as the average percentage weight variation was within the pharmacopoeial limits of 7.5%. It was found to be 0.45 to 0.75%. The weight of all the tablets was found to be uniform with low standard deviation. The mean thickness was (n=3) almost uniform in all the formulations and values ranged from 1.5 mm to 1.6 mm. The hardness values ranged from 3.9 ± 0.38 kg/cm^2 to 9.1 ± 0.24 kg/cm^2 for formulations were almost uniform.

Ibuprofen:

Table 7.9   Weight variation, thickness and hardness of ibuprofen.

Formulation code	Weight variation (%)	Tablet thickness (mm)	Hardness kg/cm2
I1	0.71%	1.6	3.6±0.36
I2	0.90%	1.5	5.2±0.34
I3	0.58%	1.6	6.0±0.34
I4	0.63%	1.5	7.4±0.35
I5	0.82%	1.5	8.0±0.30
I6	0.91%	1.6	8.6±0.22

All the tablets passed weight variation test as the average percentage weight variation was within the Pharmacopoeial limits of 7.5%. It was found to be 0.71 to 0.91%. The weight of all the tablets was found to be uniform with low standard deviation. The mean thickness was (n=3) almost uniform in all the formulations and values ranged from 5.1 mm to 4.6 mm. The hardness values ranged from 3.6 ± 0.36 kg/cm^2 to 8.6 ± 0.22 kg/cm^2 for formulations were almost uniform.

Friability and disintegration time

Table 7.10   Friability, disintegration time of aspirin and ibuprofen.

Aspirin			Ibuprofen		
Formulation code	Friability (%)	Disintegration time (minute)	Formulation code	Friability (%)	Disintegration time (minute)
A1	0.96±0.22	1.2 ± 0.27	I1	0.97±0.08	2.30 ± 0.24
A2	0.80±0.021	2.10 ± 0.23	I2	0.81±0.046	3.10 ± 0.31
A3	0.46±0.02	3.42 ± 0.18	I3	0.46±0.042	6.04 ± 0.26
A4	0.34±0.018	5.24 ± 0. 21	I4	0.34±0.051	8.00± 0.30
A5	0.28±0.014	6.0 ± 0.19	I5	0.28±0.034	10.30± 0.24
A6	0.26±0.012	8.32 ± 0.16	I6	0.26±0.021	12.03± 0.29

In vitro % drug release study

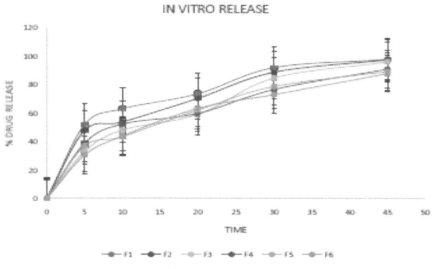

Figure 7.8   % drug release of aspirin.

The formulations A1-A6 showed more than 80% drug release. Among those six batches A1 batch showed highest drug release of 98.21%. The data for in vitro drug release of formulation was shown in Table 7.8, and in vitro drug release profile was shown in Figure 7.8.

The batch I1-I6 showed more than 80% drug release. Among those six batches I1 batch showed highest drug release of 101.04%. The data for in vitro drug release of formulation was shown in Table 7.9 and in vitro drug release profile were shown in Figure 7.9. So, all the study has indicated that tablets were prepared satisfactorily with appropriate weight variation, hardness, thickness, disintegration and friability.

It has been observed that as the concentration of the binder increases and hardness and disintegration increases with decrease in friability which may be due to the reason that with the increase in the binder concentration the binding capacity increases which leads to delay in the disintegration time and drug release and lowering friability. So as the results were satisfactory for both water soluble and insoluble drugs, *Lepidium sativum* powder can be used as a binder for BCS class I and BCS class II drugs.

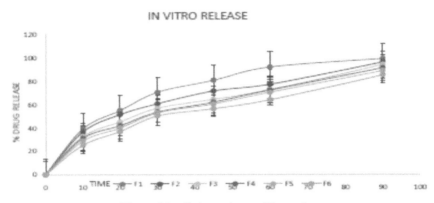

Figure 7.9   % drug release of Ibuprofen.

Comparison of standard binder with optimised batches

Precompression evaluation

Table 7.11   Bulk density tapped density Carr's Index comparison.

Formula binder weight (%w/w)	Bulk density (g/ml)	Tapped density (g/ml)	Carr's Index (%)	Hausner ratio	Angle of repose (θ)
3% (Aspirin)	0.2398	0.2697	12.46	1.12	27.16
3% (Ibuprofen)	0.2664	0.2947	10.62	1.10	27.45
3% (P.V.P.)	0.2754	0.3084	17.10	1.30	30.22

Post-compression evaluation

The results of the comparison study indicate that optimised formulation A3 and I3 have shown comparable results as compared with the results of the tablet prepared standard binder. So, it revealed that *Lepidium sativum* can be used as an alternative, cost-effective, easily available binder in pharmaceutical industry.

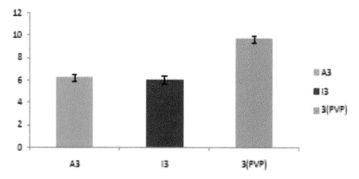

Figure 7.10 Weight variation, tablet thickness, hardness comparison.

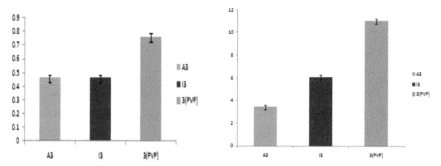

Figure 7.11 Friability and disintegration time comparison.

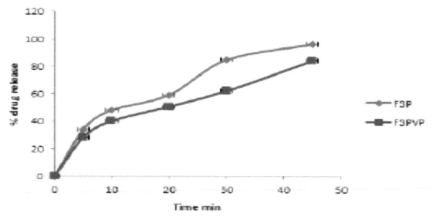

Figure 7.12 In vitro % drug release graph comparison aspirin.

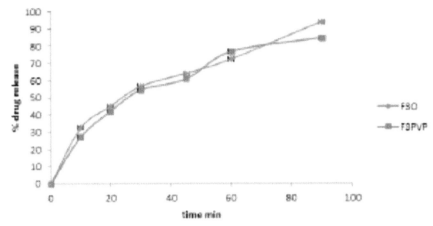

Figure 7.13   In vitro % drug release data comparison ibuprofen.

7.4 Conclusions

This study shows the effect of *Lepidium sativum* as binder in the formulation development of aspirin and ibuprofen tablets in comparison with the standard binders. From the results, it is concluded that, *Lepidium sativum* has a better binding capacity in comparison with polyvinyl pyrrolidine. The *Lepidium sativum* used is having high binding capacity, its disintegration time falls within the standard limits, the mechanical properties of the tablets were assessed using the crushing strength and friability of the tablet. Drug release properties of the tablets were assessed using disintegration time and dissolution time as assessment parameters. So it is concluded that *Lepidium sativum* can be an alternative binder for the pharmaceutical formulations. Abundant availability, food grade status, economic feasibility, commercial suitability and reliability make the mucilage as an alternative for the existing synthetic excipients.

References

[1] Davision RL. Handbook of water-soluble gums and resins. New York, McGraw Hill Book Company; 1980.

[2] Mantel CL. The water-soluble gums: Their botany source and utilization. Economic Botany, 1949; 3:3-31.

[3] A. Martin Micromereteics Physical Pharmacy. Baltimores, MD: Lippincott Williams and Wilkins; 2001:423-454.

[4] H. Liberman, L. Lachman, The Theory and practice of Industrial pharmacy, 3rd Edition, Verghese Publication House; 1991:171-193.

[5] The Indian Pharmacopoeia Delhi, The Controller of Publication, 1996;1:707.

[6] M. H. Liberman, L. Lachman, Theory and practice of Industrial pharmacy, 3rd Edition, Verghese Publication House1991;171-193.

[7] E. Aultan Pharmceutics the Science of dosage form second Edition, Churchil Livingstone, 2002;397-439.

[8] Government of India Ministry of Health & Family Welfare. Indian Pharmacopeia. Controller of Publication; 2010.

[9] US-FDA Dissolution methods for drug products 2008 http://www.access data.fda.gov/scripts/cder/dissolution/dspSearchResultsDissolutions.cfm

8

Role of Polyhydroxyalkanoates (PHA-biodegradable Polymer) in Food Packaging

**Abhishek Dutt Tripathi[1,*], Simmie Sebstraien[1],
Kamlesh Kumar Maurya[1], Suresh Kumar Srivastava[2],
Shankar Khade[2] and Kundan[2]**

[1]Centre of Food Science and Technology, Institute of Agricultural Sciences,
Banaras Hindu University, Uttar Pradesh, India
[2]School of Biochemical Engineering, Indian Institute of Technology (BHU),
Varanasi, Uttar Pradesh, India
E-mail: abhi_itbhu80@rediffmail.com
*Corresponding Author

Bioplastics are more ecofriendly than petrochemical derived plastics due their environmental degradability.PHA based plastic easily biodegradable in soil beneath natural circumstance. Bioplastics are not cause any type of hazardous pollution. Biodegradable plastic used in various industries such as medical, agricultural and packaging industry. These bioplastics component like PHA and PHB produces by various microorganism using agro industrial wastage. So it is more convenient over petrochemical packaging material and very less emission of carbon.

8.1 Introduction

Petrochemical-derived plastics being used in daily items and industries like clothing, housing, construction, furniture, automobiles, household items, agriculture, horticulture, irrigation, packaging, medical appliances; electronics and electrical, etc. drive the demand growth. According to a recent survey, the production of plastic materials in 2015 was 269 million tonnes (Source: Plastics Europe (PEMRG)/Consultic), which shows the high demand for plastics. Plastics play a very important role in our everyday life being it is used in

135

wrapping, conveniently carrying, as a protective covering, etc.. So eradication of plastics from the system is not at all possible even though it poses a great threat to our society and environment. Petrochemically derived plastics are nonbiodegradable and hence cannot be degraded over time and hence dumped in landfills, Air pollution as the only way to get rid of this plastics are to burn them which produces dangerous gases that are a source of air pollution; water pollution due to dumping and accumulation of plastics into water bodies. To overcome all these ill effects of petrochemically derived plastics to humanity and environment there was needed to work on new perspectives like biopolymers. The only solution to reduce plastic residues is the use of biodegradable plastics [43]. Biopolymers are considered as an important innovation of sustainable development as they reduce the emission of carbon dioxide and also conserve the fossil resources. Also now the manufacturing of conventional plastics has become increasingly expensive due to depletion of crude oil source. New paths are being approached for the production of biodegradable plastics by deriving it from biopolymers using starch, sugar or cellulose to substitute conventional plastics.

Initially, natural biopolymers were being used for plastic production, such as wool, leather, silk and cellulose which has been now replaced by the synthetic plastic materials but now advance technologies are being used for the production of biopolymers. Most commercialised biopolymer which has been recently much investigated is poly hydroxyl alkanoates (PHAs). PHAs the biodegradable, natural, biocompatible and renewable biopolymer. They are usually found intercellularly in bacteria when they are subjected to stressful conditions which can be nitrogen, phosphate, or oxygen limitation. Excess of carbon source and non-availability of optimum pH also add to PHA production [36]. Type of micro-organisms and nature of carbon source decides the composition of PHAs. A wide range of microorganisms produces, viz. *Ralstonia eutropha* (formerly called *Alcaligenes eutrophus, Wautersia eutropha,* or *Cupriavidus necator*), *Alcaligenes latus, Aeromonas hydrophila, Pseudomonas putida* and recombinant *Escherichia coli* [95] but it was also observed that some bacteria such as *Ralstonia eutropha* and *Alcaligenes latus* could synthesis PHAs using nutrient non-limiting media [76] unlike the one that requires limiting media for PHA production. PHAs are classified by the chain length of the branching polymers or in simpler words we can say by the number of carbon atom present in the monomers. They are classified into Short chain length (SCL) PHAs which are composed of 3-5 carbon atoms, while medium-chain-length (mcl-PHAs) and long chain- length (LCl-PHAs) consist of 6-14 and over 14 carbon atoms, respectively. Such classification arises as PHA synthesis is highly substrate specific and accept only 3-hydroxyalkanoates (3HAs) having certain carbon length [3, 4].

Poly 3-hydroxybutyrate (PHB) and polyhydroxyvalerate (PHV) are the most common form of PHA found in microbial cells. They are very much similar to the petrochemical plastics and hence can be used as an alternative to the conventional plastics [14]. In the mid-1920s, Le Moigne at the Pasteur Institute in Paris firstly identified the presence of PHB in *Bacillus megaterium*. Poly3-hydroxybutyrate (PHB) is a homopolymer of 3 hydroxybutyrates. PHB can be accumulated up to 80% of the cell dry weight from various carbon sources by *Ralstonia eutropha* [50] and near 90% in recombinant *E. coli* [115]. Blending PHAs with other monomers and copolymer production has increased the scope for new inventions in this field. PHAs are Snow being widely used in the food, agriculture, and biomedical materials [14]. PHAs are considered to be an alternative to the conventional plastics because of the similarities they possess such as high melting temperature (175°C) and relatively high tensile strength (30–35 M Pa) while the elongation at break is very different between PHB (5%) and PP (400%) [29]. PHB is free of heterogeneities, like catalyst residues or other impurities due to its natural origin.

Table 8.1   List of Limiting Compounds Leading To PHA Formation.

Compound	Organism
Ammonium	*Alcaligenes eutrophus, A. latus, Azospirillum brasiliense, Pseudomonas oleovorans, P. cepacia, Rhodospirillum rubrum, R. sphaeroides, Methylocystis parvus, Rhizobium ORS 571*
Carbon	*Spirillum sp., Hyphomicrobium sp.*
Iron	*Pseudomonas sp. K*
Magnesium	*Pseudomonas sp. K, Pseudomonas oleovorans, Rhizobium ORS 571*
Oxygen	*Azospirillum brasiliense, Azotobacter vinelandii, A. beijerinckii, Rhizobium ORS 571*
Phosphate	*Rhodospirillum rubrum, Rhodobacter sphaeroides, Caulobacter crescentus, Pseudomonas oleovorans*
Potassium, Sulphate	*Bacillus thuringiensis, Pseudomonas sp. K, Pseudomonas oleovorans, Rhodospirillum rubrum, Rhodobacter sphaeroides*

Nowadays, the research work is being emphasised on the production of PHAs by utilisation of cheap and renewable raw materials because of the relatively high production cost in comparison to that of petroleum-based plastics [14]. The requirement of special growth medium, expensive raw materials, and high recovery cost add up to the limitations of PHAs production. Carbon source normally costs up to 50% of the production cost. Hence the best alternative for this is the agro-industrial wastes and by-products such as olive oil mill effluents, sugarcane molasses, or paper mill wastewater rather than refined organic substrates [35] which can be used as carbon source.

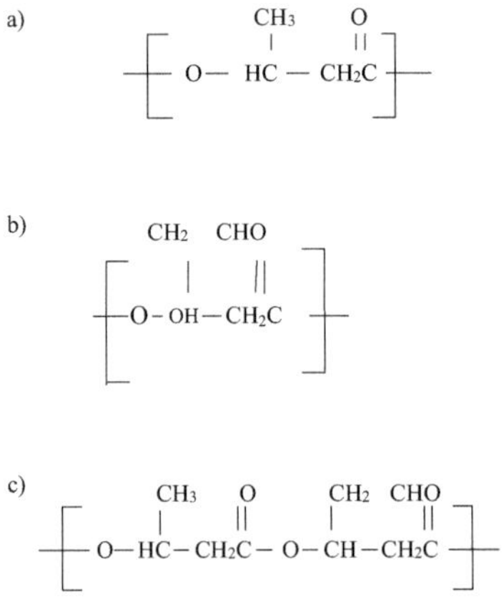

a)

$$\left[O- HC - CH_2C \right]$$
$$\text{CH}_3 \quad \text{O}$$

b)

$$\left[O-OH-CH_2C \right]$$
$$\text{CH}_2 \quad \text{CHO}$$

c)

$$\left[O-HC-CH_2C-O-CH-CH_2C \right]$$
$$\text{CH}_3 \quad \text{O} \quad \text{CH}_2 \quad \text{CHO}$$

Figure 8.1   Chemical structure of some polyhydroxyalkanoates a) PH3B b) PHV c) PHBV.

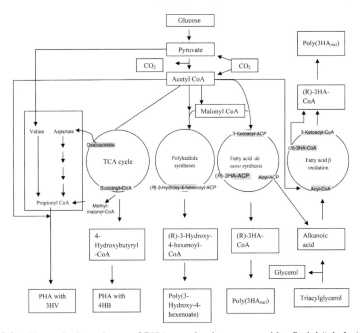

Figure 8.2   Biosynthetic pathway of PHAs production proposed by Steinbüchel, A. and B. Füchtenbusch [96].

8.2 Production

It has been reported that over 250 different bacteria can accumulate PHAs which consists of both gram-negative as well as gram-positive species. (Steinbüchel, 1991; Lenz et al., 1992). PHAs are found as an intercellular accumulated product as discrete granules in cells of Bactria, having various size and number per cell depending on the different species. It was observed in *Alcaligenes eutrophus* about 8 to 13 granules per cell having diameter range of 0.2 to 0.5μm, when observed under electron microscope the granules appeared as highly refractive inclusion (Byrom, 1994)

Intercellular depolymerases can degrade the accumulated PHA and when the nutrient limiting condition is restored this PHAs are used up as carbon and as a source of energy. The molecular weight of polymers of PHAs ranges from 2×105 to 3×106 Daltons, which are dependent on the type of microorganism and the condition provided to the microorganism for growth (Byrom, 1994). PHAs are composed of R(-)-3-hydroxy alkanoic acid monomers having carbon atoms from C4 to C14. It can be saturated or unsaturated and can be a straight or branched chain and also have side chain consisting of the aliphatic or aromatic compound (Doi et al., 1992; De Smet et al., 1983). Sudan black or Nile blue dye can be used to identify the PHA accumulating microorganisms [5, 88].

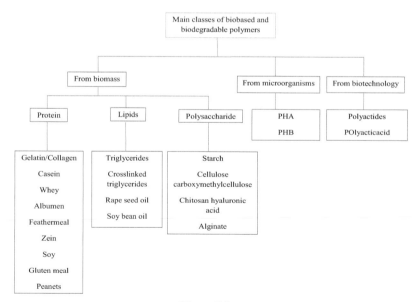

Figure 8.3

As described earlier number of carbon atom in monomer unit are divided into the short chain length polyhydroxyalkanoates PHAs, and medium chain length polyhydroxyalkanoates PHAs. *A. eutrophus* can only polymerise 3Has (SCL) while that of *Pseudomonas oleovorans* 3HAs (MCL) can be polymerised for the synthesis of PHAs.

For the synthesis of PHAs (SCL), the monomer units are oxidised at positions other than the third carbons while for PHAs(MCL), all the monomers units are oxidised except that of the third position with few exception (Valentin et al., 1994).

Industrial production of PHAs

Industrial production of PHAs consists of microbial fermentation for which carbohydrates such as glucose and sucrose are preferred as raw material. Researchers are going on for the production of development of transgenic plants which can produce PHA and can be stored in their tissues [27]. Some of the works being done in this field are taking non-food competing sources for the cultivation of bacteria in case of cyanobacteria in olive mill wastewater and their genetic modification [18] and from the municipal wastewater.

Table 8.2   Available brands of PHAs in the market.

Commercial name	Producer	Country	Product
Biomer	Biomer	Germany	Biomer P209 Biomer P226 Biomer P240
Minerv-PHA	Bio-on	Italy	MINERV-PHA^{TM}
Biogreen	Mitsubishi Gas	Japan	Biogreen
Biocycle	PHB Industrial	Brazil	BIOCYCLE 1000 BIOCYCLE 18BC-1 BIOCYCLE 189C-1 BIOCYCLE 189D-1
Ecogen	Tianan Biological Material Polyone	China	ENMAT Y1000 ENMAT Y1000P ENMAT Y3000 ENMAT Y3000P

(Continued)

Table 8.2 Continued.

Commercial name	Producer	Country	Product
Mirel	Metabolix	USA	Mirel P4001 Mirel P4010 Mirel P5001 Mirel P5004 Mirel M2100 Mirel M2200 Mirel M4100
Nodax	P&G Chemicals	USA/Japan	Nodax^{TM}
Metabolix	Telles LLC	USA	Mvera^{TM} B5011 Mvera^{TM} B5010
Jiangsu Nantian	Jiangsu Nantian Group	China	P(3HB)
Goodfellow	Goodfellow Cambridge Ltd	UK	Polyhydroxyalkanoate – Biopolymer (PHA) Polyhydroxybutyrate/ Polyhydroxyvalerate 12% - Biopolymer (PHB88/PHV12)
Tepha	Tepha Inc	USA	P(4HB)

Table 8.3 The accumulation of PHAs in various microorganisms.

Genus	PHA% (in dry biomass)	Substrate for PHA production
Acinetobacter	<1	Glucose
Aphanothecae	<1	-
Azospirillum	57	3-hydroxybutyrate
Axobacter	73	Glucose
Bacillus	25	Glucose
Beggiatoa	57	Acetate
Beijerinckia	38	Glucose
Caulobacter	36	Glucose/Glutamate/Yeast extract
Chloroflexus	<1	Yeast extract/Glycylglycine
Chlorogloea	10	Acetate/CO_2
Chromatium	20	Acetate
Chromobacterium	37	Glucose/Peptone

(*Continued*)

Table 8.3   Continued.

Genus	PHA% (in dry biomass)	Substrate for PHA production
Clostridium	13	Tryptone/Peptone/Glucose
Derxia	26	Glucose
Halobacterium	38	Glucose
Leptothrix	67	Pyruvate
Methylobacterium	47	Methanol
Methylocystis	70	Methane
Methylosinus	25	Methane
Micrococcus	28	Peptone/Tryptone
Nocardia	14	Butane
Pseudomonas	67	Methanol
Ralstonia	96	Glucose
Rhodobacter	60	Acetate
Rhodospirillum	47	Acetate
Sphaerotilus	45	Glucose/Peptone
Spirillum	40	Lactate
Spirulina	6	CO_2
Streptomyces	4	Glucose
Syntrophomonas	30	Crotonate

PHAs production from molasses

Molasses is one of the most abundantly found by-product of sugar industry and is a sugar-rich substrate that can be used for the production of PHAs. Molasses finds a massive use at an industrial level as carbon source due to its low price and its wide availability [2]. Production of PHB was first reported by Page by *Azotobacter vinelandii* UWD using sugar beet molasses [77]. PHAs accumulation was also reported in media derived from soy molasses; it was also tried to produce mcl-PHA from soy molasses using *Pseudomonas corrugate* [93].

PHAs production from whey and whey hydrolysates

Whey is a by-product generated from the dairy industry and more specifically from the manufacture of cheese. The carbon source in whey is the lactose which constitutes the 70% of the total dry weight which is widely used in fermentative industry. Recombinant E. coli strains were constructed which were able to express *Cupriavidus necator* phaC2 gene (also known as *Ralstonia eutropha*) for the production of PHB from whey [58]. By controlling the oxygen supply, the concentration of intercellular PHB was increased in recombinant E. coli. Increase in PHB up to 70-80% was observed when the aeration rate was kept constant but the maximum agitation speed was at 500 rpm, without removing the culture broth [51].

PHAs production from lignocellulosic raw materials

Lignocellulosic is non-food biomass which can be used as a raw material for the synthesis of PHB as it is considered as a most abundant renewable carbon source [60]. The compositional constituents of lignocellulose are cellulose (30-50%), hemicelluloses (20-50%) and lignin (15-35%). It has been observed that some bacterium such as *Saccharophagus degradans* ATCC 43961 were able to directly utilise cellulose for PHA [70]. Lignocellulosic biomass is utilised for the conversion of monomers of sugar from hemicelluloses such as xylose, arabinose, mannose, galactose and rhamnose. Recalcitrance nature of lignocellulose is one of the major drawbacks for the usage as raw material as it requires pre-treatment steps to produce bioplastics. Production of PHB/V from fructose and glucose can also be achieved using propionic acid as a co-substrate [80, 81]. In order to obtain xylose utilising strains for the PHAs production, researchers isolated and screened 55 bacteria from soil [91] in which two strains, *Burkholderia cepacia* IPT048 and *B. sacchari* IPT 101 showed good PHB synthesis abilities.

PHAs production from fats, vegetable oils and waste cooking oils

For fermentative PHA production triacylglycerides (TAG) and its derived fatty acids were used in the 1990s as they have more energy per mole when compared to carbohydrates for conversion in PHA synthesis [93]. To avoid the saponification step, it is desirable to use the triacylglycerols in fatty acids production. Saturated PHA monomers were produced when a high level of saturated fats was used from coconut oil while unsaturated PHA monomers were produced when a high level of unsaturated fats from soybean oil was used. Now researchers are looking for the production of PHA from various oils such as palm oil [112], olive oil [73], corn oil [13], coconut oil [99], soybean oil [37, 42], other vegetable oils and animal fats. Due to the recent increase in edible oil now PHA production from oils are not considered as cost-effective.

PHAs production from glycerol

Glycerol is the major by-product of the biodiesel. *Pseudomonas putida* KT2442 was shown to produce mcl-PHA from glycerol [93]. The polymers produced were similar to that of the one produced from glucose or fructose as a carbon source [40]. Work was done in the field of constructing a recombinant *E. coli* strain with the phaC1 gene from *Pseudomonas* sp. LDC-5 [98]. This recombinant strain produced 3.4 g PHAs/L on glycerol and fish peptone derived medium. Recently, researcher studied PHB biosynthesis using glycerol as a carbon source [22] while some studied the different uses of crude glycerol in the production of PHA [68] and also the influence of the salt concentration in the glycerol on the PHB production by *C. necator* JMP134 and *Paracoccus denitrificans* [12].

PHAs production from wastewater

PHA can be produced from wastewater which can reduce the cost of raw material, viz. municipal wastewater [17, 19, 20], biodiesel wastewater [26], food processing waste effluent [48, 106], brewery waste effluent [61], paper mill wastewater [6] and kraft mill wastewater [79] which are source of organic water, have been tested for PHAs biosynthesis. The first step involved the conversion of organic carbon from source to volatile fatty acids and then into PHA by mixing cell cultures [6, 61].

Production of biopolymer

The PHAs which are prominent having high concentration with high productivity are PHB, poly (3-hydroxybutyrate-co-3-hydroxyvalerate) and poly (3-hydroxyhexanoate-co-3- hydroxyoctanoate). Different strategies are being introduced for the efficient production of PHAs (Yu, 2001, Du et al.,2001; Du and Yu, 2002). The efficient production of PHA can only be obtained by considering different attributes for example selection of the microbes which can take up carbohydrate source which is relatively cheaper and easily available, should have high polymer synthesis rate and have maximum polymer accumulation capacity. To improve production methods such as fed-batch and continuous cultivation are being implemented (Lee, 1996; Du and Yu, 2002a; Du and Yu, 2002b; Du et al. 2001b; Yu and Wang, 2001). Researchers have derived some equations which can relate to the PHA yield on different carbon sources. (Yamane, 1992; Yamane, 1993; Yu and Wang, 2001).

Table 8.4   List of microorganisms producing PHA on different biomass.

Biomass	Microorganism	PHA Content, Cell dry weight (%CDW)
Soy molasses	*Pseudomonas corrugata*	5.00-17.00
Molasses (sugarcane)	*Bacillus cereus SPV*	61.07
Crude glycerol	*Cupriavidus necator DSM545*	62.70
Wheat straw hydrolysate	*Burkholderia sacchari DSM17165*	72.00
Extruded rice bran and wheat bran	*Haloferax mediterranei*	56.00
Rice grain distillery wastewater with nutrient added	*Activated sludge (MMC)*	67.00
Olive mill wastewater	*Cupriavidus necator DSM 545*	55.00 (11% 3HV co-polymer)
Waste rapeseed oil	*Cupriavidus necator H16*	76.00 (8% 3-HV co-polymer)
Pineapple juice	*Bacillus sp.*	48.89
Sugarbeet juice	*Azohydromonas lata*	65.60
Maple sap	*Azohydromonas lata*	77.60
Oil palm frond juice	*Cupriavidus necator NCIMB 11599*	75.00

8.3 Characterisation and Identification

Apart from an economical process for the production of PHAs, it is also important to characterise the isolated PHA. The quantitative estimation and characterisation of PHAs, can be done using various techniques which include staining reactions, spectrophotometric analysis, infrared-FTIR spectroscopy, HPLC, flow cytometry and spectrofluorometry, gas chromatography-GCMS, NMR spectroscopy, molecular weight determination and thermal analysis of the extracted polymer.

8.3.1 Spectrophotometric Methods

This method is helpful in the quantitative estimation of the esters extracted using solvent extraction method. The application of this method is limited to the case when the quantity of polyesters is relatively high. Researchers have developed an analytical method which consists of involves measuring the turbidity of P(3HB) granules produced by the digestion of bacterial cells with sodium hypochlorite solution [111]. As the P(3HB) molecule gets degraded at high temperature, different spectrophotometric methods were developed. The P(3HB) gets degraded into crotonic acid when heated in the presence of concentrated sulphuric acid and using crotonic acid Ultraviolet (UV) absorption band at an absorbance of 235 nm the polyester content can be determined [56, 110].

8.3.2 Infrared Spectroscopy

The main principle behind using this method is that P (3HB) molecule has a strong carbonyl absorption peak at 1728 cm^{-1} in the infrared spectrum. It was reported that the presence of B-hydroxybutyric acid in intact bacterial cells using infrared spectrophotometry [8]. Fourier transform infrared spectroscopy method can be employed to analyze the chemical structure of the extracted PHA. Biopolymer was dissolved in chloroform and using KBr disc (Shamala et al., 2003) in Perkin Elmer Fourier transform infrared (FTIR) spectrophotometer (Jasco FTIR6100, Japan) infrared spectra of the samples were taken at the range of 400-4000 cm^{-1}. Using FTIR the functional group of PHB granules were predicted, i.e., aliphatic CHO stretching, =C–H deformation, =C–H, =CH and =C–O [75]. It was observed that the maximum PHA concentration of 70.8% (CDW) in *Alcaligenes* sp. by supplementing palmitic acid while the yield increased to 78.0% (DCW) and productivity of 0.14 g l^{-1} h in a shake flask incubation. FTIR technique can be used to determine the synthesis of PHA [78] and also useful in the investigation of copolymers (SCL–MCL PHA) and also the different compositional structure. It was observed that a band at 3434 cm^{-1} shows the presence of OH alcohol or OH carboxyl polymeric of the polymer compound and a band having absorption at 1648 cm^{-1} indicates the presence of the ketoenolic group. With variation in copolymer composition, the band of the CO stretching vibration also changed [63]. Bruker model IFS-55 FTIR spectrometer coupled to a Bruker IR microscope fitted with an IBM compatible PC running OPUS, Version 2.2 software was used to prepare a sample for FTIR and spectra was taken at 500 Hz [101].

8.3.3 High-Performance Liquid Chromatography (HPLC)

High pressure liquid chromatography (HPLC) was employed to measure the yield of poly-B-hydroxybutyrate (PHB) in *Rhizobium japonicum* bactericides [45]. The products in the acid digest of PHB containing material were fractionated by HPLC on Aminex HPX-87-H ion exclusion resin for organic acid analysis. The degradation of PHB into crotonic acid, when subjected to acid digestion, showed absorbance at 210 nm.

8.3.4 Gas Chromatography-Mass Spectrometry (GC-MS)

This method was firstly used for the detection of poly-3-hydroxybutyrate [10]. In this process, the cells were treated with mild acid or alkaline hydrolysis which is followed by GC. This process was improvised by using propanol and HCl [85]. Researchers have done works which described trimethylsilyl (TMSi) derivatisations and have a wide application in the determination of PHA [57].

8.3.5 NMR Spectroscopy

Using nuclear magnetic resonance (NMR) spectra [7, 28, 33, 47], the copolymers of PHA can be determined. NMR spectroscopy is more efficient because in this process the hydrolysis step can be skipped. This method is to check the accumulation of PHB isolated from samples of *Pseudomonas sp.* Strain LBr by using cross polarisation magic-angle spinning 13C NMR spectra [41].

8.3.6 Flow cytometry and Spectrofluorometry

Flow cytometry has been used in *A. Eutrophus* which showed the advantage of single cell light scattering measurement for analysis and control of fermentation processes [94]. This method along with cell sorting technology has immense application in strain improvement. This method was also advocated for characterisation and differentiation of bacteria during microbial processes [69]. The relation between total culture fluorescence and total biomass concentration was also shown in work performed by workers [32]. Using Nile Red fluorescence technique researchers developed a convenient method for PHB detection [24]. Nile blue was used as a stain for E. coli in flow cytometry. (Jose et al., 2009).

8.3.7 Staining Reactions and Microscopy

Different staining techniques are being employed for detection of bacterial strains which can detect PHA producing bacteria. Sudan black

(Gerhardt et al., 1981), Nile blue [5] Nile red (Gorenflo et al., 1999) stains are able to detect intercellular PHA in bacterial strains. Quantitative analysis of PHA can be done using spectrophotometric method [56], gas chromatography and HPLC (Brandl et al., 1988). Sudan black B stain is used to detect the presence of PHB while phase contrast microscopy can also be used which is beneficial when large numbers of isolates are to be screened [9]. Sudan black B was used as a dye for the isolation and detection of P (3HB) producing colonies on a nitrogen limiting plate [88]. Nile blue is a water soluble organic dye of basic nature (Anthony et al., 1982). PHB granules have bright orange fluorescence when observed under a microscope with an episcopic-fluorescence attachment. This dye is more efficient as it doesn't stain any other inclusion bodies.

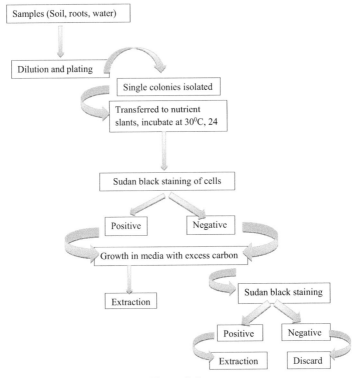

Figure 8.4

8.4 Extraction and Recovery

Extraction of PHA can be very tedious and can be economically expensive hence efforts are made to develop methods with can make the downstream process easier to recover fermented product.

8.4.1 Using Chloroform and Sodium Hypochlorite

PHA extraction was done using chloroform–hypochlorite extraction method and pure PHA can be extracted using nonsolvent precipitation and then filtration done using membrane filter (2 mm, Millipore). The nonsolvents that can be used are methanol and water (7:3, vol-1) (Srivastava and Tripathi, 2012). Recovery of PHA can be made with a dispersion of hypochlorite solution and chloroform. By centrifugation of PHA producing cell mass with hypochlorite and chloroform solution provide a three separate phase. The bottom-most chloroform layer is containing PHA which can be recovered using a pipette [34].

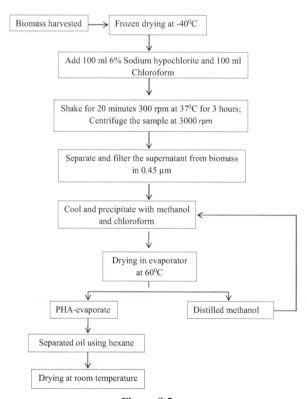

Figure 8.5

8.4.2 Using Surfactant and Chelating Agents

Centrifuged cell suspended in a known amount of water with 0.6% Triton and 0.06% EDTA adjusted at a pH of 13 was rested at 50°C for 10 min and again centrifuged. The precipitates containing PHA further purified by dissolving in chloroform and precipitated with hexane (Chen et al., 2001). It is reported that a PHA-containing biomass which was treated with SDS or Triton X-100 keeping ratio 1:1, had recovery of PHA with purity of 87% however, when SDS or Triton X-100 treatment further followed by hypochlorite wash, purity up to 97 to 98% achieved [82]. In the mixed surfactant system when 0.1 M ammonium chloride was added at pH 3, maximum extraction of PHA achieved with recovery of 84.4% and a purity of 92.49% at a reduced cloud point temperature of 33°C [71]. Extraction of PHA by disruption of cell membranes using fatty acid carboxylates as surfactants, provide excellent polymer recovery (>99%) and high purity (>90%) [86]. Yang et al. [113] reported most of the detergents which are surfactant or mixture of surfactant recovered PHA polymer in high purity from cells. Alkylbenzene sulfonic acid (LAS-99) in linear form which is a detergent produced a high yield of highly pure PHAs, and less amount of detergent needed as compared to SDS to produce comparable yields. Chemical extraction of PHA with detergents could potentially substitute use harsh organic solvents. Thus it leads to making cleaner technology process for industrial production of PHA.

8.4.3 Using Alkali

Centrifuged cell suspended in a known amount of water and pH adjusted to 11 rested at 50°C for 10 min was again centrifuged. The precipitate hence found was purified by dissolving in chloroform and precipitated with hexane (Choi and Lee, 1999). Biomass (dried) containing poly(3-hydroxybutyrate) an important PHA was treated with 5 different chemicals (sodium hydroxide [NaOH, 1 M]), sodium hypochlorite (NaClO, 100 g/L), SDS(100 g/L), EDTA(50 g/L) or sulphuric acid (H_2SO_4, 100 g/L). These reaction mixtures were then thermostated at 50^0C for 10 min at 125 rpm. Biomass was separated from the aqueous fraction and precipitate containing mostly PHB washed with distilled water twice and then dried at 40°C in the oven for 24 h. and the purity of PHB was analyzed. Corresponding digestion chemicals of highest purity of PHA were chosen as the second-stage digestion-chemicals. Among different methods used, sodium dodecyl sulphate (SDS)–NaOH method was selected as the best method [62]. For poly R-hydroxybutyrate recovery disruption of bacterium *Ralstonia eutropha* cells by supercritical CO_2. For pretreatment, the cells were exposed to 140 mM NaCl and heat (60°C, 1 h) and cells were also exposed to 0.2-0.8% (w/w) NaOH to examine the effect of alkaline pretreatment. It was found that pretreatment with a minimum of 0.4% (w/w) NaOH was necessary [49].

8.4.4 Using Enzymes

This is a separate process which had the capacity in the commercial production of PHB by BIOPOL [39]. Optimisation of the action of this enzyme mixture was also done (Choi and Lee 1999). In case of *Rhizobium meliloti* 14, the recovery was up to 90% (Lakshman and Shamla 2003). Cell biomass of *P. putida* re-suspended in water and sterilised. Suspension digested with alcalase and SDS at pH 8.5 and at 55°C followed by further treatments with EDTA and lysozyme at pH 7 at 30°C for 15 min. To recover the PHA granules in water suspension solubilised non-PHA removed by cross-flow ultrafiltration system and purified through continuous diafiltration process. Final purity of PHA in water suspension obtained 92.6%, with a nearly 90% recovery [114]. PHB producing *Ralstonia eutropha* DSM545 cells were suspended in 25 ml (concentration of 20 g/L) in a specific buffer depending on the enzyme to be used; 1ml of different enzyme suspension of bovine chymotrypsin, bovine trypsin, bromelain, papain, lysozyme bovine and pancreatin was added separately to achieve the desired enzyme mass per biomass. Under desired temperature solutions were agitated at 200 rpm. Samples were diluted to 1:50 with distilled water. The best cell lysis was achieved with 2.0% of bromelain, equivalent to 14.1 Uml^{-1}, at 50°C and pH 9.0, with 88.8% PHB purity [44]. Medium chain length PHAs producing *fluorescent Pseudomonas*, *P. oleovorans, P. putida* GPp104 and *P. putida* KT2442 cells were harvested in the late stationary phase. Cells were exposed to various lytic treatments. The solubilisation involves heat, protease and detergent; leaves the peptidoglycan intact; and facilitates the separation. The enzymes alcalase, neutrase and lecitase were used. By solubilisation, the purity of the resulting medium chain length PHAs exceeds 95% [23].

8.4.5 Using Microbial Method of Extraction

The lytic enzymes are able to dissolve the cell material which helps in the release of accumulated PHA to the broth which can be extracted from broth using a minimal amount of chloroform. This is one of the most effective methods in which lytic enzyme of an actinomycetes culture was used to lyse *Rhizobium meliloti* 14 cells (Lakshman and Shamla 2003). In the presence of sucrose as carbon substrate, *Sinorhizobium meliloti* produced polyhydroxyalkanoate (PHA) which constitute 50% of the biomass. Using secondary fermentation involving a cell lytic actinomycetes species *Microbispora* sp. the intracellular PHA was isolated. Up to further 72 hr, *S. meliloti* fermented broth kept at 30°C, 150 rpm with supplementation. Pelleted growth of *Microbispora* sp. cells was removed by filtration and filtrate was extracted by a surfactant and a chelating agent. Maximum yield of PHA obtained was 49% of biomass weight after 24 h of lytic culture fermentation recovery of the polymer was 94% with 90% purity [55].

8.4.6 Purification of Biopolymers

An efficient recovery method should have high purity yield and should be cost-effective. The conventional approach includes different methods, viz. chemical digestion, enzymatic digestion, solvent extraction, by using a surfactant and mechanical. For PHA recovery, organic solvents and chemicals are being used from prolong time. Extraction of PHAs with organic solvents is found to be better than other extraction methods [38]. In solvent extraction process, improved cellular membrane permeability and solubilisation recover PHA without degradation [83]. Contamination of Gram-negative bacterial endotoxin of the polymer is prevented by this method, and thus, improved polymeric quality is utilised for biomedical applications (Lee et al.1999). It was found that solvents that contain functional carbon atoms with one hydrogen atom and one chlorine at least atom give better results. Extraction with diols, acetalised triols, dicarboxylic or tricarboxylic acid esters and butyrolactone could result in high purity recovery [100]. Under physical condition 34.5°C temperature, 6.54 pH and agitation speed of 3.13 Hz *Alcaligenes sp.* cultivated for the production of PHB. Extraction of PHB performed using 50 mL chloroform hypochlorite dispersion extraction. Achieved PHB mass fraction yield of 76.80% on dry molasses substrate [104].

Another commonly used method is chemical digestion. For differential digestion of non-PHA biomass, sodium hypochlorite was used. It was reported that using sodium hypochlorite solution and chloroform, in *C. necator* and recombinant *E. coli* high purity levels of PHA obtained [34]. Alkali treatment was commonly given to get higher recovery. NaOH has been used in various concentrations. Acidic treatment has not gained importance. Chelating agents like EDTA play an important role in recovery.

Surfactants integrate itself into the lipid bilayer membrane of the cell and thus increase the cell envelope volume till saturation. As result cells lysed and production of micelles of surfactant and membrane phospholipids took place, which releases P(3HB) having cell debris in its surrounding [82]. Anionic sodium dodecyl sulphate (SDS) is one of the most common surfactants. Combine effect of chelate and surfactant could enhance PHA release. Recovery of P(3HB) from *C. necator* was 93.3% with 98.7% purity was obtained [16].

Zeneca Group PLC had developed enzymatic digestion method [39]. Recover of medium-chain-length (mcl) PHA from *P. putida* with enzymatic digestion has achieved purity of almost 93% [46]. Various enzymes, viz. lysozyme, neutrase, alcalase, lecitase, etc. have been exploited. For the extraction of PHA, enzymes with hypochlorite, SDS or alkali treatment, EDTA, the heat were also tested [23].

Mechanical cell disruption is also a method for PHAs recovery. This method includes bead mill disruption, high-pressure homogenisation ultrasonication and centrifugation techniques. To recover P(3HB) from *Methylobacterium* sp., V49, 5% (w/v) SDS solution was used and homogenised at operating pressure of 400 kg/cm^2. The yield of 98% and purity of 95% were obtained after two cycles [31]. P(3HB) recovery of 80% and a purity of 98.5% (w/w) were achieved with chemical treatment followed by three cycles of centrifugation to remove non-P(3HB) cell material [105].

Biological recovery method is an alternative to conventional methods of recovery. This method is safe and eco-friendly. Kunasundari et al. [54] have proposed a recovery process using laboratory rats. Their study reported certain animals could feed *C. necator* H16 cells as a protein source. Indigestible PHA granules digested excreted out in white faecal pellets form have almost 90% PHA by weight by rinsing with water resulted in an increased PHA purity of greater than 95%. Under the publication number WO2010134798 A1 this method has been patented. This approach is natural and environmentally friendly; it exploited the digestive system of suitable animals. It requires no chemical and has specific requirements so cost wise it is very effective. Enzymes in the digestive tract of animals can hydrolyze the biomass and the compounds such as PHA which cannot be digested, excreted as faecal pellets [97]. Although faeces produced, require multiple washing. Washings could be on the basis of percentage of PHA in the faeces. PHA recovered biologically may be seen beyond bad feeling which may endorse its applicability in products.

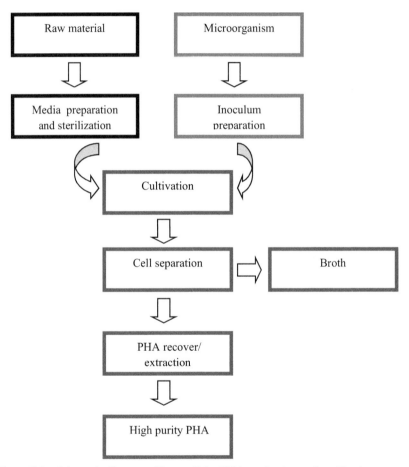

Figure 8.6   Schematic diagram of intracellular PHA production and purification process.

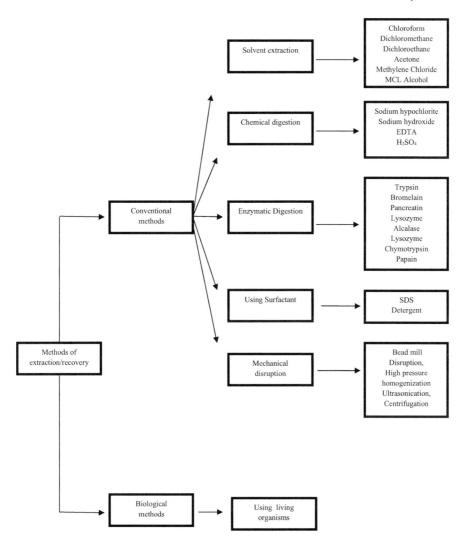

Figure 8.7 Conventional and biological recovery methods of PHA from bacterial cells.

8.4.7 Application of Biopolymers in Food Packaging

Treatment of plastic waste by thermal conversion or landfill disposal has dire ecological consequences. Incineration generates toxic products and growing plastic waste heaps like problems are of major concern. Plastic recycling rate around the world is 20% only this cannot be a potential solution. An approach for bio-based plastic materials can bring a solution. Now it has a market less than 5% of the entire plastic market. Several plastic-like materials available in the market are thermoplasticised starch, polylactic acid, bio-based polyethylene, poly trimethylene terephthalate, polybutylenes succinate, poly(p-phenylene), microbial poly hydroxyl alkanoates (PHAs) [15]. Among these only PHAs are completely bio-synthesised and having complete biocompatibility and biodegradability [15, 107].

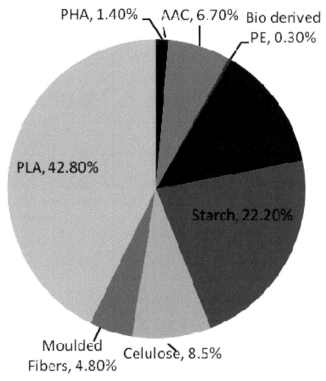

Figure 8.8   Global bioplastic packaging market by product type (%utilisation) 2010.

AAC: aliphatic-aromatic copolyesters; PLA: polylactic acid; PHB: polyhydroxyalkanoate; WSP: water-soluble polymers
Source: Pira International Ltd

Any food packaging materials should have properties of protecting food from the external threats like contaminants, radiation, mechanical damage, humidity and also should be maintaining organoleptic properties of food. Maintaining a modified atmosphere inside the package, stability under low temperature, the environmental conditions should not degrade it during the storage of the food product, but the material must degrade after discarding it.

Aspects of PHAs which makes it food packaging material:

- PHAs can be converted into films solely or in the composite form *via* thermoforming [108].
- Due to adaptable crystallinity and elasticity, PHAs can be processed to flexible foils or rigid and strong cast components (*via* injection moulding) for making storage boxes and containers [67, 92].
- Hydrophobicity of PHA films exhibit a high water vapour barrier; parallelly high barrier properties against CO_2 [21, 66].
- *scl*-PHA copolyesters (PHBV), homopolyesters PHB and poly(4HB) exhibit a high oxygen barrier. The fine-tuning of the polyester composition during the biosynthesis determines gas barrier performance [65].
- Their high UV-barrier protects food spoilage [92].
- With food residues, PHA-based packaging spoiled can be easily discarded by composting [52].

A problem associated with PHAs that endotoxins attached to PHAs produced from Gram-negative strains [30]. Also, lipid residues remain attached after extraction can affect sensory quality of food it encloses. It is reported that endotoxins can be removed using supercritical solvent extraction of PHAs. Chemical methods can remove remaining lipids and pyrolytic, also by using a mixture of organic solvents and anti-solvent or high-pressure extraction with anti-solvents [53].

It is important to have oxygen, water and CO_2 barrier in packaging material. To maintain product packaging material should overcome the driving force of the difference in oxygen partial pressures inside and outside of the package. Water and oxygen barrier of PHAs comparable to polystyrene, another common petrochemical non-biodegradable and biodegradable packaging material, poly ε-caprolactone (PCL) [87].

Table 8.5   Water and oxygen barrier properties of PHA and other polymers [14].

	PHA	LDPE	HDPE	PET	Nylon
Water	5–19	1.2	0.5	1.3	25
Oxygen	23–29	250–840	30–250	5	3

Water vapour transmission in g-mil (100 cm^2/day) at 38°C, 90% RH
Oxygen transfer rate in cc-mil (100cm^2/day) at 25°C 0% RH
PHB was reported as having a good barrier against CO_2 permeation [21, 89, 92]. Also, the effect of addition of saturated and unsaturated fatty acid on the production of PHB was analyzed and palmitic acid found to be best among other saturated and unsaturated fatty acid to give better production [103]. PHA biopolyesters meet required features necessary for biodegradable food packaging material. For production, advanced and inexpensive methods are already developed. For large-scale PHA biosynthesis inexpensive raw materials or CO_2, upgraded to function as carbon sources. Tripathi et al. [102] have reported production of PHAs by soil bacterium *Pseudomonas aeruginosa* using sugar refinery waste (cane molasses) produced the maximum PHA (biodegradable polymer) among another carbon source.

Table 8.6   Molecular weight comparison of PHB obtained from Various sources.

Compound	M_n	M_w	M_n/M_w	Viscosity (*n*)
PHB extracted from *Alcaligenes* sp.	1.25×10^5	2.86×10^5	2.29	9.8×10^1
PHB (*A. latus*)	3.13×10^5	5.28×10^5	1.69	9.4×10^1
PHB (commercial)	0.91×10^5	1.77×10^5	1.95	9.1×10^1
PCL (Polycaprolactone)	0.56×10^5	1.63×10^5	2.90	8.5×10^1
PHB (Soy)	3.48×10^5	7.90×10^5	2.26	8.5×10^1

Mw: Weight-average molecular weight; Mn: Number-average molecular weight

Table 8.7   Thermo-mechanical features of different PHAs (Koller 2014).

	PHB	Poly-(3HB-co-3%-3HV)	Poly-(3HB-co-11%-3HV)	Poly-(3HB-co-20%-3HV)	Poly-(3HB-co-3%-4HB)	Poly-(3HB-co-16%-4HB)	Poly-(3HB-co-64%-4HB)	Poly-(4HB)	Poly-(3HB-co-21,8-%-3HV-co-5,1-%-4HB)	Poly-(3HB-co-2,4-%-3HV-co-13,4-%-3HHx)	Poly-(3HO-co-12%-3HHx)
Melting temperature [°C]	177	170	157	145	166	152	50	60	140	114	61
Glass transition temperature [°C]	4	-	2	-1	-	-8	-	-50	-2	-2	-35
Tensile strength [MPa]	40	38	38	32	28	26	17	104	-	-	9
Elongation at break [%]	6	-	5	50	45	444	591	1000	-	481	380

HHx: hydroxyhexanoate; HB: hydroxybutyrate; HV: hydroxyvalerate; HO: hydroxyoctanoate

8.5 Biodegradability

All PHA family members are biodegradable in the soil, in freshwater and seawater and in industrial composting equipment. It is very versatile family; these can be processed to hard or soft plastics and to crystalline and amorphous polymers as well. American Society for Testing & Materials (ASTM) has set criteria for a plastic to be called compostable. Compostable plastic needs to be met following criteria:

Biodegrade - Break down into carbon dioxide, water, and biomass.
Disintegrate - Material is indistinguishable in the compost.
Eco-toxicity - Should not produce any toxic material and can support plant growth.

Degradability of PHA materials:

a) Environmental degradability

PHA materials are biodegradable in various environments. Temperature, moisture level, pH, and nutrient supply affect rate of biodegradation of PHA materials. Composition, crystallinity, additives, and surface area of the polymer also important factors in determining the rate of biodegradation of PHA. Various bacteria and fungi can degrade PHA. Microorganisms excrete extracellular PHA-degrading enzymes to break down the solid polymer into water soluble oligomers and monomer, so that utilise it as nutrients within cells. PHB has an important property that it can be completely broken down to CO_2 and H_2O by microorganisms. Many publications approve that PHB and its composites can be degraded in natural ecosystems, viz. soil, compost and water. Maergaert et al. isolated from soil more than 300 microbial strains capable of degrading PHB in vitro.

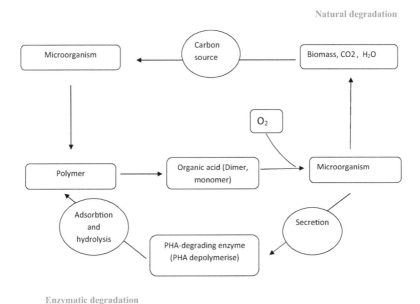

Figure 8.9   Biodegradation process of PHA in a natural environment.

8.5.1 Enzymatic Degradability

Biodegradation PHAs depends on the secretion of extracellular PHA depolymerases, which are carboxylesterases (EC 3.1.1.75 and EC 3.1.1.76), and on the physical state of PHAs. It is found that N-terminal of structural genes of extracellular PHB depolymerases have catalytic domain whereas C-terminal have a substrate-binding domain, and both domains connected and maintained by optimal distance with linker region. Analysis reveals that C-terminal domain has substrate-binding domain for water-insoluble P(3HB) substrate, and PHB depolymerase lost hydrolyzing activity without C-terminal domain for water-insoluble P(3HB) but retained activity for water-soluble (R)-3HB oligomers [72]. The rate of enzymatic activity is highly dependent on the enzyme concentration. With the concentration of PHB depolymerase, rates of enzymatic activity increased to a certain value and later gradually decrease [27]. The rates of film erosion (biodegradation) of P(3HB-*co*-10 mol%-6HH) samples with 40–50% crystallinity were higher than those of P(3HB) samples with 63–78% crystallinity [1]. The enzymatic degradation of P(3HB) homopolymer and five types of copolyesters were performed in aqueous solutions of purified PHA depolymerase secreted from *R. pickettii* T1 at 37°C. It was reported that rate of enzymatic erosion on the solution cast increase with an increase in the fraction of second monomer units up to 10–20 mol% then it showed a decrease in the erosion rate [74].

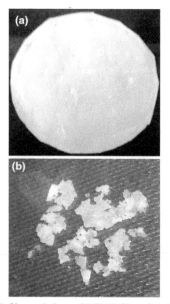

(a) PHB film at 0 days, (b) PHB film after 50 days

Figure 8.10 Degradation of the PHB film in garden soil [19].

Figure 8.11 PHA bottles disintegrate in the soil within 60 days (but remain intact until discarded).

Source: https://www.biobasedpress.eu/2016/08/pha-promising-versatile-biodegradable/

ISO 17556, ASTM D5988, NF U52-001 and UNI 11462 are the international standards for testing biodegradability of plastics in the soil. These standards are specifically designed to determine plastics biodegradability in soil under natural conditions. Pass levels and time frame are required for a bio-based product to ensure biodegradability under soil conditions.

8.5.2 Versions and Title of Standard Testing Methods for Determining Biodegradability of Materials in Soil

8.5.3 American Society for Testing and Materials International (ASTM)

ASTM D 5988-12: Standard test method for determining aerobic biodegradation of plastic materials in soil

International Standards
ISO 17556-2012: Determination of the ultimate aerobic biodegradability of plastic materials in soil by measuring the oxygen demand in a respirometer or the amount of carbon dioxide evolved
ISO 11266-1994: Soil quality - Guidance on laboratory testing for biodegradation of organic chemicals in soil under aerobic conditions

8.5.4 French and Italian Normalisation Organisations (AFNOR, UNI)

NF U52-001 February 2005: Biodegradable materials for use in agriculture and horticulture - Mulching products -Requirements and test methods

UNI 11462:2012: Plastic materials biodegradable in soil - Types, requirements and test methods

8.5.5 OECD Guidelines

304A: Inherent Biodegradability in Soil
307: Aerobic & Anaerobic Transformations in Soil

8.6 Challenges and Opportunity

Bioplastics are favourable and more preferred over conventional petro derived plastics regarding their eco-friendly nature as today's major concern is the pollution caused by these non-degradable plastics. They have wide application in various fields, viz. medical, agricultural and packaging industry. Unlike the petro based plastics, bioplastics do not produce effluent to the surrounding nature, hence safer to use. The problem being faced in case of synthetic plastic is the lack of waste management system that is a great threat to humankind in the long run. These all aspects lead to the research and development in the field of bioplastics.

As the fossil fuels are on the verge of depletion, hence there is a need to develop alternatives to synthetic plastics. The production of PHA can be done using substrates that are waste from any food industry or even municipal wastewater solids can be used for the accumulation of PHAs. Ethylene is a hazardous gas is produced by the synthetic plastics, can be avoided using bioplastics.

There are wide scope and opportunity in finding the most suitable and efficient strain of microorganism which produces PHA, developing a suitable source of the better production of PHA, looking for the favourable, cheap and reliable source of raw material and substrate which is rich in carbon source. While choosing the raw material or the recovery of end product the economical aspect needs to be considered (Scott, 2000).

References

[1] Abe, H., Y. Doi, et al. "Solid-state structures and enzymatic degradabilities for melt-crystallized films of copolymers of (R)-3-hydroxybutyric acid with different hydroxyalkanoic acids." Macromolecules 1998; 31(6): 1791-1797.

[2] Albuquerque MGE, Eiroa M, Torres C, Nunes BR, Reis Mam. Strategies for the Development of a Side Stream Process for Polyhydroxyalkanoate (PHA) Production From Sugar Cane Molasses. J Biotechnol 2007; 130: 411-21.

[3] Aldor IS, Keasling JD. Process Design for Microbial Plastic Factories: Metabolic Engineering of Spolyhydroxyalkanoates. Curropin Biotech 2003; 14: 475-83.

[4] Anderson AJ, Dawes EA. Occurrence, Metabolism, Metabolic Role, And Industrial Uses Of Bacterial Polyhydroxyalkanoates. Microbiol Rev 1990; 54: 450-72

[5] Anthony-G. Ostle and J.G. Holt. "Nile blue A as a fluorescent stain for poly-B-hydroxybutyrate". Applied and environmental Microbiology. 44 (1982): 238-241.

[6] Bengtsson S, Werker A, Christensson M, Welander T. Production Of Polyhydroxyalkanoates by Activated Sludge Treating A Paper Mill Wastewater. Bioresource Technol 2008; 99: 509-16.

[7] Bioembergen S, Holden DA, Hamer GK, Bluhm TL and Marchessault RH "Studies on composition and crystallinity of bacterial poly (B-hydroxybutyrate-co-B-hydroxyvalerate)". Macromolecules. 19(1986):2865-2871.

[8] Blackwood AC and Agnes EPP "Identification of B-hydroxybutyric acid in bacterial cells by infrared spectroscopy". J. Bact. 74 (1957):266-267.

[9] Bourque B., Ouellette, G. Andre, D. Groleau. "Production of poly-B-hydroxybutyrate from Methanol: Characterization of a new isolate. *Methylobacterium extorquens*". Applied Microbiol. Biotechnol. 37(1992): 7-12.

[10] Brauneeg G, Sonnleitner B and Lafferty RM "A rapid gas chromatographic method for the determination of Poly-B-hydroxybutyric acid in microbial biomass". Eur. J. Appl. Microbiol Biotechnol. 6(1978): 29-37.

[11] Bugnicourt, E., P. Cinelli, et al. (2014). "Polyhydroxyalkanoate (PHA): Review of synthesis, characteristics, processing and potential applications in packaging."

[12] Cavalheiro JMBT, de Almeida MCMD, Grandfils C, da Fonseca MMR. Poly(3-Hydroxybutyrate) Production by *Cupriavidus necator* Using Waste Glycerol. Process Biochem 2009; 44: 509-15.

[13] Chaudhry W, Jamil N, Ali I, Ayaz M, Hasnain S. Screening For Polyhydroxyalkanoate (PHA)-Producing Bacterial Strains And Comparison

Of PHA Production From Various Inexpensive Carbon Sources. Ann Microbiol 2011; 61: 6239.

[14] Chen G-Q. A Microbial Polyhydroxyalkanoates (PHA) Based Bio and Materials Industry. Chemsoc Rev 2009; 38: 2434-46.

[15] Chen, G.-Q. (2010). Introduction of Bacterial Plastics PHA, PLA, PBS, PE, PTT, and PPP. Plastics from Bacteria, Springer: 1-16.

[16] Chen, Y., J. Chen, et al. (1999). "Recovery of poly-3-hydroxybutyrate from Alcaligenes eutrophus by surfactant–chelate aqueous system." Process Biochemistry 34(2): 153-157.

[17] Chua ASM, Takabatake H, Satoh H, Mino T. Production Of Poly-hydroxyalkanoates (PHA) By Activated Sludge Treating Municipal Wastewater: Effect Of pH, Sludge Retention Time (SRT), and Acetate Concentration In Influent. Water Res 2003; 37: 3602-11.

[18] Cinelli P., Lazzeri A., Anguillesi I., Bugnicourt E. OLI-PHA A Novel And Efficient Method For The Production Of Polyhydroxyalkanoate Polymer-Based Packaging From Olive Oil Waste Water. In 'Proceed-ings Of 3rd International Conference On Industrial And Hazardous Waste Management. Chania, Crete, Greece' P8 (2012).

[19] Coats ER, Loge FJ, Wolcott MP, Englund K, Mcdonald AG. Synthesis Of Polyhydroxyalkanoates In Municipal Wastewater Treatment Water Environ Res 2007; 79: 2396-403.

[20] Coats ER, Vandevoort KE, Darby JL, Loge FJ. Toward Polyhydrox-yalkanoate Production Concurrent With Municipal Wastewater Treat-ment In A Sequencing Batch Reactor System. J Environ Eng 2011; 137: 46-54.

[21] Dagnon KL, C. Thellen, et al. (2010). "Physical and thermal analysis of the degradation of poly (3-hydroxybutyrate-co-4-hydroxybutyrate) coated paper in a constructed soil medium." *Journal of Polymers and the Environment* 18(4): 510-522.

[22] De Almeida A, Nikel PI, Giordano AM, Pettinari MJ. Effects Of Granule-Associated Protein Phap On Glycerol-Dependent Growth And Polymer Production In Poly(3-Hydroxybutyrate)-Producing *Escherichia coli*. Appl Environ Microb 2007; 73: 7912-6.

[23] De Koning, G. and B. Witholt (1997). "A process for the recovery of poly (hydroxyalkanoates) from Pseudomonads Part 1: Solubilization." Bioprocess and Biosystems Engineering 17(1): 7-13.

[24] Degelau A, Scheper T. Bailey J.E. Guske C. "Fluorometric Mea-surement of Poly-B-hydroxybutyrate in *Alcaligenes eutrophus* by

flow cytometry and spectrofluorometry". Appl. Microbiol. Biotechnol. 42(1995): 653-657.

[25] Dirk Betscheider Joachim Jose, "Nile blue A for staining *E. coli* in flow cytometry experiments". Analytical Biochemistry, 384.1(2009) :194-196.

[26] Dobroth ZT, Hu S, Coats ER, Mcdonald AG. Polyhydroxybutyrate Synthesis On Biodiesel Wastewater Using Mixed Microbial Consortia. Bioresource Technol 2011; 102: 3352-9.

[27] Doi Y., Mukai K., Kasuya K., Yamada K.: Biodegradation Of Biosynthetic And Chemosynthetic Polyhydroxyalkanoates. Studies In Polymer Science, 12, 39–51 (1994).

[28] Doi, Y. Kunioka M., Nakamura Y. and Soga K. "Nuclear magnetic resonance studies on unusual bacterial copolyesters of 3-hydroxybutyrate and 4-hydroxybutyrate". Macromolecules. 21(1988): 2722-2727.

[29] El-Hadi A., Schnabel R., Straube E., Müller G., Henning S.: Correlation Between Degree Of Crystallinity, Morphology, Glass Temperature, Mechanical Properties And Biodegradation Of Poly (3-Hydroxyalkanoate) Phas And Their Blends. Polymer Testing, 21, 665–674 (2002). Doi: 10.1016/S0142-9418(01)00142-8

[30] Furrer, P., S. Panke, et al. (2007). "Efficient recovery of low endotoxin medium-chain-length poly ([R]-3-hydroxyalkanoate) from bacterial biomass." Journal of microbiological methods 69(1): 206-213.

[31] Ghatnekar, M.S., J.S. Pai, et al. (2002). "Production and recovery of poly−3−hydroxybutyrate from Methylobacterium sp. V49." Journal of chemical technology and biotechnology 77(4): 444-448.

[32] Groom C.A., Luong J.H.T. and Mulchandani A., "Online culture fluorescence measurement during the batch cultivation of poly-β-hydroxybutyrate producing *Alcaligenes eutrophus*." J. Biotechnology 8(1988): 271-278.

[33] Gross R.A., C. Demello, R.W. Lenz, Brandl H. and Fuller R.C. "Biosynthesis and characterization of poly (B-hydroxyalkanoates) produced by *Pseudomonas oleoverans*". Macromolecules. 22(1989): 1106-1115.

[34] Hahn, S. K., Y. K. Chang, et al. (1994). "Optimization of microbial poly (3−hydroxybutyrate) recover using dispersions of sodium hypochlorite solution and chloroform." Biotechnology and Bioengineering 44(2): 256-261.

[35] Halami Pm. Production Of Polyhydroxyalkanoate From Starch By The Native Isolate *bacillus Cereus* Cfr 06. *World J Microbiolbiotechnol.* 2008; 24: 805812.

[36] Hazer B, Steinbuchel A. Increased Diversification Of Polyhydrox-yalkanoates By Modification Reactions For Industrial And Medical Applications. *Appl Microbiol Biotechnol.* 2007; 74: 112.

[37] He W, Tian W, Zhang G, Chen G-Q, Zhang Z. Production Of Novel Polyhydroxyalkanoates By *Pseudomonas stutzeri* 1317 From Glucose And Soybean Oil. Fems Microbiol Lett 1998; 169: 45-9.

[38] Heng, K.-S & Ong, Su Yean & Sudesh, Kumar. (2016). Efficient biosynthesis and recovery of polyhydroxyalkanoate. 12. 383-398.

[39] Holmes, P. A. and G. B. Lim (1990). Separation process, Google Patents

[40] Huijberts GNM, Eggink G, De Waard P, Huisman GW, Witholt B. *Pseudomonas putida* KT2442 Cultivated On Glucose Accumulates Poly(3-Hydroxyalkanoates) Consisting Of Saturated And Unsaturated Monomers. Appl Environ Microb 1992; 58: 536-44.

[41] Jacob G.S., Garbow J.R. and Schaefer J. "Direct measurement of poly (B-hydroxybutyrate) in *Pseudomonas* by solid state 13C NMR". J. Biol. Chem. 261 (1986):16785-16787.

[42] Jacob G.S., Garbow J.R. and Schaefer J. "Direct measurement of poly (B-hydroxybutyrate) in *Pseudomonas* by solid state 13C NMR". J. Biol. Chem. 261 (1986):16785-16787.

[43] Kahar P, Tsuge T, Taguchi K, Doi Y. High Yield Production Of Poly-hydroxyalkanoates From Soybean Oil By *Ralstonia eutropha* And Its Recombinant Strain. Polym Degrad Stabil 2004; 83: 79-86.

[44] Kalia VC, Chauhan A, Bhattacharyya G, Rashmi P. Genomic Databases Yield Novel Bioplastic Producers. *Nat Biotechnol.* 2003; 21:845846.

[45] Kapritchkoff, F.M., A.P. Viotti, et al. (2006). "Enzymatic recovery and purification of polyhydroxybutyrate produced by Ralstonia eutropha." Journal of Biotechnology 122(4): 453-462.

[46] Karr D.B., Waters J.K. and Emerich D.W., "Analysis of Poly-B-Hydroxybutyrate in *Rhizobium japonicum* bacterioids by ion-exclusion high pressure liquid chromatography and UV detection". Appl. Env. Microbiol. 46.6(1983): 1339-1344.

[47] Kathiraser, Y., M.K. Aroua, et al. (2007). "Chemical characterization of medium−chain−length polyhydroxyalkanoates (PHAs) recovered by

enzymatic treatment and ultrafiltration." Journal of Chemical Technology and Biotechnology 82(9): 847-855.

[48] Kathryn P. Caballero, Steven F. Karel, Richard A Register. "Biosynthesis and Characterization of Hydroxybuyurate copolymers". International Journal of Biological Macromolecules 17.2(1995): 86-92

[49] Khardenavis AA, Suresh Kumar M, Mudliar SN, Chakrabarti T. Biotechnological Conversion Of Agro-Industrial Wastewaters Into Biodegradable Plastic, Poly β-Hydroxybutyrate. Bioresource Technol 2007; 98: 3579-84.

[50] Khosravi−Darani, K., E. Vasheghani−Farahani, et al. (2004). "Effect of process variables on supercritical fluid disruption of Ralstonia eutropha cells for poly (R−hydroxybutyrate) recovery." Biotechnology Progress 20(6): 1757-1765.

[51] Kim BS, Lee SC, Lee SY, Chang HN, Chang YK, Woo SI. Production Of Poly(3-Hydroxybutyric-Co-3-Hydroxyvaleric Acid) By Fedbatch Culture Of *Alcaligeneseutrophus* with Substrate Control Using On-Line Glucose Analyzer. Enzyme Microb Tech 1994; 16: 556-61.

[52] Kim Bs. Production Of Poly(3-Hydroxybutyrate) From Inexpensive Substrates. Enzyme Microb Tech 2000; 27: 774-7.

[53] Koller, M., A. Salerno, et al. (2010). "Modern biotechnological polymer synthesis: a review." Food Technology and Biotechnology 48(3): 255-269.

[54] Koller, M. (2014). "Poly (hydroxyalkanoates) for food packaging: Application and attempts towards implementation." Applied Food Biotechnology 1(1): 3-15.

[55] Kunasundari, B., V. Murugaiyah, et al. (2013). "Revisiting the single cell protein application of Cupriavidus necator H16 and recovering bioplastic granules simultaneously." PLoS ONE 8(10): e78528.

[56] Lakshman, K. and T.R. Shamala (2006). "Extraction of polyhydroxyalkanoate from Sinorhizobium meliloti cells using Microbispora sp. culture and its enzymes." Enzyme and Microbial Technology 39(7): 1471-1475.

[57] Law J.H. and R.A. Slepecky, "Assay of Poly-B- Hydroxybutyric acid". J. Bacteriol. 82 (1961): 33-36.

[58] Lee Eun Veol and Choi Cha Yong, "Gas chromatography Mass spectrometric analysis and its application to a screening procedure for novel bacterial polyhydroxy alkanoic acids containing long chain saturated and unsaturated monomers". J. Ferm Bioengg. 80.4(1995): 408-414.

[59] Lee SY, Middelberg APJ, Lee YK. Poly(3-Hydroxybutyrate) Production From Whey Using Recombinant *Escherichia coli*. Biotechnol Lett 1997; 19: 1033-5

[60] Lemoigne M. Products Of Dehydration And Of Polymerization Of β-Hydroxybutyric Acid. Bull Soc Chem Biol. 1926

[61] Lin CSK, Luque R, Clark JH, Webb C, Du C. Wheat-Based Biorefining Strategy For Fermentative Production And Chemical Transformations Of Succinic Acid. Biofuels BioprodBior 2012, 6: 88-104.

[62] Liu H-Y, P VandergheynstJs, Darby JL, Thompson DE, Green PG, Loge FJ. Factorial Experimental Designs For Enhancement Of Concurrent Poly(Hydroxyalkanoate) Production And Brewery Wastewater Treatment. Water Environ Res 2011; 83: 36-43.

[63] Lo, C.-W., H.-S. Wu, et al. (2011). "High throughput study of separation of poly (3-hydroxybutyrate) from recombinant *Escherichia coli XL1* blue." Journal of the Taiwan Institute of Chemical Engineers 42(2): 240-246.

[64] Luo, R., Chen, J., Zhang, L. and Chen, G. Polyhydroxyalkanoate copolyesters produced by *Ralstonia eutropha* PHB-4 harboring a low-substrate-specificity PHA synthase phaC2Ps from *Pseudomonas stutzeri* 1317. *Biochemical Engineering Journal* (2006) 32(3), 218-225.

[65] Mergaert J, Anderson C, Wouters A, Swings J, Kersters K. Biodegradation of polyhydroxyalkanoates. FEMS Microbiol Rev (2-4), 317-321 (1992).

[66] Martin, D.P. and S.F. Williams (2003). "Medical applications of poly-4-hydroxybutyrate: a strong flexible absorbable biomaterial." Biochemical Engineering Journal 16(2): 97-105.

[67] Miguel, O. and J.J. Iruin (1999). Evaluation of the transport properties of poly (3−hydroxybutyrate) and its 3−hydroxyvalerate copolymers for packaging applications. Macromolecular Symposia, Wiley Online Library.

[68] Mensitieri, G., E. Di Maio, et al. (2011). "Processing and shelf life issues of selected food packaging materials and structures from renewable resources." Trends in Food Science & Technology 22(2): 72-80.

[69] Mothes G, Schnorpfeil C, Ackermann JU. Production Of PHB From Crude Glycerol. Engineering In Life Sciences. 2007; 7: 475-9.

[70] Muller S., Losche A. Bley T, Scheper T., "A flow cytometric approach for characterization and differentiation of bacteria during microbial processes". Appl Microbiol Biotechnol 43 (1995):93-101.

[71] Muller S., Losche A. Bley T, Scheper T., "A flow cytometric approach for characterization and differentiation of bacteria during microbial processes". Appl. Microbiol. Biotechnol 43(1995):93-101.

[72] Munoz Lea, Riley Mr. Utilization Of Cellulosic Waste From Tequila Bagasse And Production Of Polyhydroxyalkanoate (Pha) Bioplastics By *Saccharophagus degradans*. Biotechnol Bioeng 2008; 100: 882-8

[73] Murugesan, S. and R. Iyyaswami (2017). "Nonionic surfactants induced cloud point extraction of Polyhydroxyalkanoate (PHA) from Cupriavidus necator." Separation Science and Technology(just-accepted).

[74] Nojiri, M. and T. Saito (1997). "Structure and function of poly (3-hydroxybutyrate) depolymerase from *Alcaligenes faecalis* T1." Journal of Bacteriology 179(22): 6965-6970.

[75] Ntaikou I, Kourmentza C, Koutrouli EC *et al.* Exploitation Of Olive Oil Mill Wastewater For Combined Biohydrogen And Biopolymers Production. Bioresource Technol 2009; 100: 3724-30.

[76] Numata, K., H. Abe, et al. (2009). "Biodegradability of poly (hydrox-yalkanoate) materials." Materials 2(3): 1104-1126.

[77] Nur ZY, Belma A, Yavuz B, Nazime M (2004) Effect of carbon and nitrogen sources and incubation time on poly-beta-hydroxybutyrate (PHB) synthesis by Bacillus megaterium 12. Afr J Biotechnol 3:63–69

[78] Ojumu TV, Yu J, Solomon BO. Production Of Polyhydroxyalkanoates-A Bacterial Biodegradable Polymer. Afr J Biotechnol 2004;3: 18-24.

[79] Page WJ. Production Of Polyhydroxyalkanoates By *Azotobacter vinelandii* UWD In Beet Molasses Culture. Fems Microbiol Rev 1992; 103: 149-57

[80] Pan, J., Li, G., Chen, Z., Zhu, W. and Xu, K. Alternative block polyurethanes based on poly(3-hydroxybutyrate-co4-hydroxybutyrate) and poly(ethylene glycol). *Biomaterials* (2009). 30 2975–2984.

[81] Pozo G, Villamar AC, Martínez M, Vidal G. Polyhydroxyalkanoates (Pha) Biosynthesis From Kraft Mill Wastewaters: Biomass Origin And C:N Relationship Influence. Water Sci Technol 2011; 63: 449-55.

[82] Ramsay BA, Lomaliza K, Chavarie C, Dubé B, Bataille P, Ramsay JA. Production Of Poly-(Beta-Hydroxybutyric-Co-Betahydroxyvaleric) Acids. Appl Environ Microb 1990; 56: 2093-8.

[83] Ramsay BA, Ramsay JA, Cooper DG. Production Of Polyhydroxyalka-noic Acid By *Pseudomonas cepacia*. Appl Environ Microb 1989; 55: 584-9.

[84] Ramsay, J., E. Berger, et al. (1990). "Recovery of poly-3-hydroxyalkanoic acid granules by a surfactant-hypochlorite treatment." Biotechnology Techniques 4(4): 221-226.

[85] Ramsay, J., E. Berger, et al. (1994). "Extraction of poly-3-hydroxybutyrate using chlorinated solvents." Biotechnology Techniques 8(8): 589-594

[86] Reddy CSK, Ghai R, Rashmi, Kalai VC. Polyhydroxyalkanoates: An Overview. Bioresour Technol 2003;87(2):137e46

[87] Riis V. and Mai W. J. "Chromatographic determination of Poly-B-hydroxybutyric acid from microbial biomass after hydrochloric acid propanolysis". Chromatogr. 445(1988): 285-289.

[88] Samorì, C., M. Basaglia, et al. (2015). "Dimethyl carbonate and switchable anionic surfactants: two effective tools for the extraction of polyhydroxyalkanoates from microbial biomass." Green Chemistry 17(2): 1047-1056.

[89] Sanchez-Garcia, M., E. Gimenez, et al. (2007). "Novel PET nanocomposites of interest in food packaging applications and comparative barrier performance with biopolyester nanocomposites." Journal of Plastic Film & Sheeting 23(2): 133-148.

[90] Schlegel H.G. Lafferty, R. and Krauss I. "The isolation of mutants not accumulating Poly-B-hydroxybutyrate". Arch. Microbiol. 71(1970):283-294.

[91] Shogren, R. (1997). "Water vapor permeability of biodegradable polymers." Journal of Polymers and the Environment 5(2): 91-95.

[92] Shrivastav, A., S.K. Mishra, et al. (2011). "Biodegradability studies of polyhydroxyalkanoate (PHA) film produced by a marine bacteria using Jatropha biodiesel byproduct as a substrate." World Journal of Microbiology and Biotechnology 27(7): 1531-1541.

[93] Silva LF, Taciro MK, Michelin Ramos ME, Carter JM, Pradella JGC, Gomez JGC. Poly-3-Hydroxybutyrate (P3HB) Production By Bacteria From Xylose, Glucose And Sugarcane Bagasse Hydrolysate. J Ind Microbiol Biot 2004; 31: 245-54.

[94] Siracusa, V., P. Rocculi, et al. (2008). "Biodegradable polymers for food packaging: a review." Trends in Food Science & Technology 19(12): 634-643.

[95] Solaiman DKY, Ashby RD, Foglia TA, Marmer WN. Conversion Of Agricultural Feedstock And Coproducts Into Poly(Hydroxyalkanoates). Appl Microbiol Biot 2006; 71: 783-9.

[96] Srienc B. Arnold and Bailey J.E. "Characterization of intracellular poly-B-hydroxybutyrate (PHB) in individual cells of *Alcaligenes eutrophus* H16 by flow cytometry". Biotechnol. Bioengg 26 (1984): 982-987.

[97] Steinbüchel A, Hustede E, Liebergesell M, Pieper U, Thimma M, Valentin H. Molecular Basis For Biosynthesis And Accumulation Of polyhydroxyalkanoic Acids In Bacteria. Fems Microbiol Lett 1992;103: 217-30.

[98] Steinbüchel, A. and B. Füchtenbusch (1998). "Bacterial and other biological systems for polyester production." Trends in Biotechnology 16(10): 419-427.

[99] Sudesh, K. (2010). Extraction and purification of polyester granules. International Patent No.WO2010134798 A1.

[100] Sujatha K, Shenbagarathai R. A Study On Medium Chain Length polyhydroxyalkanoate Accumulation In *Escherichia coli* Harbouring PhaC1 Gene Of Indigenous *Pseudomonas* sp. Ldc-5. Lett Appl Microbiol 2006; 43: 607-14

[101] Thakor N, Trivedi U, Patel Kc. Biosynthesis Of Medium Chain Length Poly(3-Hydroxyalkanoates) (Mcl-Phas) By *Comamonas testosteroni* During Cultivation On Vegetable Oils. Bioresour Technol 2005; 96: 1843-50.

[102] Traussnig, H., E. Kloimstein, et al. (1990). Extracting agents for poly-D (-)-3-hydroxybutyric acid, Google Patents.

[103] Tripathi AD, Srivastava SK (2011) Kinetic study of biopolymer (PHB) synthesis in Alcaligenes sp. in submerged fermentation process using TEM. J Polym Sci Environ (2011) 19:732–738.

[104] Tripathi, A.D., A. Yadav, et al. (2012). "Utilizing of sugar refinery waste (cane molasses) for production of bio-plastic under submerged fermentation process." Journal of Polymers and the Environment 20(2): 446-453.

[105] Srivastava, S. and A.D. Tripathi (2013). "Effect of saturated and unsaturated fatty acid supplementation on bio-plastic production under submerged fermentation." 3 Biotech 3(5): 389-397.

[106] Tripathi, A. D., S. K. Srivastava, et al. (2013). "Statistical optimization of physical process variables for bio-plastic (PHB) production by Alcaligenes sp." Biomass and Bioenergy 55: 243-250.

[107] Van Wegen, R., Y. Ling, et al. (1998). "Industrial Production of Polyhydroxyalkanoates Using *Escherichia coli*: An Economic Analysis." Chemical Engineering Research and Design 76(3): 417-426.

[108] Venkateswar Reddy M, Venkata Mohan S. Influence Of Aerobic And Anoxic Microenvironments On Polyhydroxyalkanoates (PHA) Production From Food Waste And Acidogenic Effluents Using Aerobic Consortia. Bioresour Technol 2012; 103: 313-21.

[109] Verlinden, R.A., D.J. Hill, et al. (2007). "Bacterial synthesis of biodegradable polyhydroxyalkanoates." Journal of applied microbiology 102(6): 1437-1449.

[110] Virov, P. (2013). "Polyhydroxyalkanoates: Biodegradable polymers and plastics from renewable resources." Materiali in Tehnologije 47(1): 5-12.

[111] Wang, G., A.S. Lee, et al. (1999). "Accelerated solvent extraction and gas chromatography/mass spectrometry for determination of polycyclic aromatic hydrocarbons in smoked food samples." Journal of Agricultural and Food Chemistry 47(3): 1062-1066.

[112] Ward A.C. and Dawes E. "A disc assay for poly-B-hydroxybutyrate". Anal. Biochem. 52(1973):607-613.

[113] Williamson D.H. and Wilkinson J.F. "The isolation and Estimation of the Poly-B-hydroxybutyrate inclusions of *Bacillus* Species". J. Gen. Microbiol. 19(1958):198-209.

[114] Wu Ty, Mohammad Aw, JahimJm, Anuar N. A Holistic Approach To Managing Palm Oil Mill Effluent (Pome): Biotechnological Advances In The Sustainable Reuse Of Pome. Biotechnol Adv. 2009; 27: 40-52.

[115] Yang, Y.-H., C. Brigham, et al. (2011). "Improved detergent-based recovery of polyhydroxyalkanoates (PHAs)." Biotechnology letters 33(5): 937-942.

[116] Yasotha, K., M. Aroua, et al. (2006). "Recovery of medium-chain-length polyhydroxyalkanoates (PHAs) through enzymatic digestion treatments and ultrafiltration." Biochemical Engineering Journal 30(3): 260-268.

[117] Yu J, Stahl H. Microbial Utilization And Biopolyester Synthesis Of Bagasse Hydrolysates. Bioresour Technol 2008; 99: 8042-8.

9

Xylitol: Fermentative Production and Statistical Optimization Using Novel Isolates of *Candida parapsilosis* Strain BKR1 in the Indigenously Designed Multiphase Reactor

Balakrishnaraja Rengaraju[1], D. Vinotha[1] and P. Ramalingam[2]

[1]Department of Biotechnology, Bannari Amman Institute of Technology, Sathyamangalam, India
[2]Department of Biotechnology, Kumaraguru College of Technology, Coimbatore, India
E-mail: balakrishnarajar@bitsathy.ac.in

Xylitol is a natural polyol and is most widely known for its sugar substitute properties in diabetic patients; it is also used against an oral bacterial population. Most economical approach for the commercial production of xylitol involves the suitable yeast fermentation. In this present investigation, factorial optimisation of these medium and process conditions is considered. Xylitol production by *Candida parapsilosis* strain BKR1 using Plackett–Burman and RSM is reported in modified minimal medium. Medium design using Plackett–Burman screening reports that the significant components are xylose, yeast extract, potassium dihydrogen phosphate and magnesium sulphate. Further the statistical optimisation using face-centred central composite design reveals the optimal grades of the significant medium components as, xylose $\sim$104 g/l, yeast extract $\sim$4.5 g/l; $KH_2PO_4 \sim$2.8 g/l and $MgSO_4 \cdot 7H_2O \sim$2 g/l. Among the significant fermentation parameters, agitation, pH, temperature and inoculum level were optimised and validated as 107 rpm, 5.0, 30°C and 1 ml, respectively. *Candida parapsilosis* BKR1 was optimised for higher xylitol productivity using statistical approaches.

9.1 Introduction

Xylitol is one of the most expensive polyol sweeteners and considerable attention in the food and pharmaceutical industries. Potential healthcare benefits of xylitol includes, viz. tooth decay, ear infection in children, substitute for sugar in diabetic patients and parenteral application to mentally distressed patients [13, 23]. Due to its resistance towards Maillard reactions, the xylitol becomes a prospective food ingredient and continuously it has been explored because of properties such as storage, colour and taste addition to food products. Chewing gums, hard caramels, liquorice sweets, wafer fillings, chocolate and other sugar based confectioneries for diabetics [1]. Xylitol derived from detoxified substrate, can be metabolized in the absence of insulin and can replace sugar on a weight basis (w/w) [15] making it a suitable sweetener for *Diabetes mellitus* patients [2, 5]. Bioconversion of xylitol is influenced by numerous factors in various concentrations of ingredients in culture medium, so their optimization study is very significant. Response surface methodology (RSM) is a statistical analysis that is useful for modelling and optimization of medium conditions and process parameters [3]. Factorial design or response surface methodology has been utilized extensively for optimizing different biotechnological processes [21]. Several chemical steps are involved in the purification of xylose [4]. The microbial conversion of xylose to xylitol is particularly attractive in that the process is relatively easy and does not depend upon any toxic catalyst [20]. Xylitol production through bioconversion has been proposed to alternative process utilizing microorganism such as yeast, bacteria and fungi [16]. Among those, yeast has some desirable properties and was proven to be a potential xylitol producer [12, 19]. In the present chapter, the microbial screening and optimization of medium composition, process variables for xylitol production by *Candida parapsilosis* strain BKR1 using Plackett-Burman and Response Surface Methodology (RSM) are detailed. The Plackett-Burman screening design is applied for knowing the most significant nutrients enhancing xylitol production. Face-centred central composite design (FC-CCD) and Box-Benhem design were applied to determine the optimum level of each of the significant nutrients and process variables respectively. For enhanced production of highly demanding polyalcohol, xylitol in more economical way through biotechnological methods, optimization of medium components and fermentation parameters in adapted strain of *Candida parapsilosis* strain BKR1 becomes inevitable.

The present study covers the fermentative production and optimization including,

1. The indigenous yeast isolate, *Candida parapsilosis* strain BKR1 are optimized in minimal medium for higher xylitol production
2. Plackett–Burman optimization was carried out to identify significant medium components.
3. Optimization techniques such as central composite design (CCD) for medium components & Box Behnkem method (BBM) for process parameters were assessed based on response surface plots for desirable xylitol yield.

9.2 Fermentation and Statistical Optimization

The yeast strain *Candida parapsilosis* strain BKR1 (NCBI Accession No: KC462059) was isolated as described earlier [17] and the strains were maintained at 4°C in culture medium supplemented with 20 g agar. The medium composition (g/l) is given as: malt extract - 3.0; yeast extract - 3.0; peptone - 5.0; glucose - 10.0, with pH 7. It was stored at 4°C as agar slants and sub-cultured every thirty days to maintain viability. Composition of the modified minimal medium:- Ammonium sulphate – 3 g/l; KH_2PO_4 – 5 g/l; Yeast extract – 3 g/l; Magnesium sulphate ($MgSO_4$) – 0.2 g/l; Zinc Sulphate ($ZnSO_4$) – 6 mg/l; Copper Sulphate ($CuSO_4$) – 0.4 mg/l; Ferric Chloride ($FeCl_3$) – 15 mg/l; Cobalt Chloride ($CoCl_2$) – 0.45 mg/l; Manganese Sulphide ($MnSO_3$) – 5 mg/l; Calcium Chloride ($CaCl_2$) – 200 mg/l; Xylose – 100 g/l. Optimization studies were carried out in the minimal medium as described elsewhere [17].

Fermentation was carried out in 250 ml Erlenmeyer flasks with 100 ml of modified minimal medium under laboratory incubation conditions. This is supplemented with different nutrient concentration for tests according to the selected factorial design and sterilized at 121°C for 20 min as per standard autoclaving procedure. The flasks were inoculated with 1 ml of grown culture broth containing *Candida parapsilosis* strain BKR1 once the temperature reach ambient conditions aseptically under laminar hood. The flasks were maintained at 30° C under agitation at 100 rpm for 48 h based on the results obtained from preliminary screening which was carried out for 5 days (data not provided).

Table 9.1 Placket–Burman design for medium components.

S No	Code	Factor(s): Medium Composition	(-) level	(+) level
1	A	Ammonium sulphate (g/l)	2	4
2	B	Pottasium dihydrogen phosphate (g/l)	3	7
3	C	Yeast extract (g/l)	2	4
4	D	Magnesium sulphate (g/l)	0.1	0.3
5	E	Zinc sulphate (mg/l)	4	8
6	F	Copper sulphate (mg/l)	0.3	0.5
7	G	Ferric chloride (mg/l)	10	20
8	H	Cobalt chloride (mg/l)	0.3	0.6
9	J	Manganese sulphite (mg/l)	3	7
10	K	Calcium chloride (mg/l)	150	250
11	L	Xylose (g/l)	50	150

The sugars and sugar alcohols were characterized by high performance liquid chromatography (HPLC) using an ion moderated partition chromatography column SHODEX SC 1011 sugar column (300 X 7.8 mm). Samples are eluted with deionized HPLC grade water at a flow rate of 0.5 ml/min at 80°C and detected with a differential refractometer (WATERS 410). [15]

9.2.1 Plackett–Burman (PB) Experimental Design

PB experimental design assumes that there are no interactions between the different variables in the given range under consideration. A linear approach is choosen and considered to be sufficient for preliminary screening with less variables under study. Plackett–Burman experimental design is a fractional factorial design and the main effects of such a design may be simply calculated as the difference between the average of measurements made at the high level (i.e. '+1') of the factor and the average of measurements at the low level (i.e. '–1').

PBD was exploited especially to determine the number of variables suggestively affect xylitol production [3]. Eleven variables (Table 9.1) were screened in 13 experimental runs and insignificant variables were eliminated in order to obtain a simpler and meaningful set of factors for fitting the model equation. The low level ('-1') and high level ('+1') of each factor are enlisted in Table 9.1. The statistical software package Design Expert 7.0.0 (Stat-ease Inc., USA) was applied to process the experimental data. Once the critical factors were spotted through iterative screening, the central composite design

(CCD) and Box-Behnkem Method (BBM) was used to obtain a quadratic model for medium components and process parameters respectively.

9.2.2 Response Surface Methodology

The face-centred central composite design and Box Behnkem design were adopted to investigate the effects of variables on their cumulative responses and subsequently in the optimization of xylitol productivity. These methods are appropriate for fitting a quadratic surface and also help to optimize the effective parameters with manageable number of experiments, as well as to analyze the interaction between the identified significant factors/ parameters. In order to determine the existence of a relationship between the factors and response variables, the unperturbed data were analyzed in a statistical manner, using regression concept to link the similarity between the variables under study. A regression design is normally employed to model a response as a mathematical function (either known or empirical) of a few continuous factors and good model parameter estimates are desired [3].

The coded values of the process parameters are illustrated by the following equation:

$$x_i = \frac{X_i - X_0}{\Delta x} \tag{9.1}$$

where x_i – coded value of the ith variable, X_i – uncoded value of the ith test variable and X_0 – uncoded value of the ith test variable at centre point. The regression analysis is performed to estimate the response function as a second order polynomial:

$$Y = \beta_0 + \sum_{i=1}^{k} \beta_i X_i + \sum_{i=1}^{k} \beta_{ii} X_i^2 + \sum_{i=1,i<j}^{k-1} \sum_{j=2}^{k} \beta_{ij} X_i X_j \tag{9.2}$$

where Y is the predicted response, β_0 constant, β_i, β_j and β_{ij} are coefficients estimated from regression. They represent the linear, quadratic and cross products of X_i and X_j on response.

9.2.3 Model Fitting and Statistical Analysis

The regression and graphical analysis with statistical significance were carried out using Design Experts 7.0.0. In order to envisage the relationship between the experimental variables and responses, the response surface and

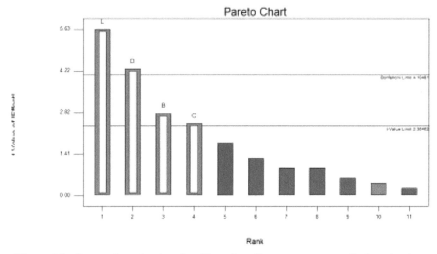

Figure 9.1 Pareto chart showing the effect of media component on xylitol production.

contour plots were generated from the models. The optimum values of the fermentation process variables were obtained from the regression equation as described elsewhere [17]. The adequacy of the models was further vindicated through analysis of variance (ANOVA). Lack-of-fit is a special diagnostic test for adequacy of a model and compares the standard error, based on the replicate measurements to the other lack of fit, based on the model performance [14]. F-value, calculated ratio between the lack-of-fit mean square and the pure error mean square are statistic parameters used to determine whether the lack-of-fit is significant or not, at a significance level. The statistical models were authenticated with respect to xylitol production under the conditions predicted by the model in laboratory shake-flasks level. Samples were drawn at the desired intervals of fermentation time and xylitol production was determined as described above using characterisation methods.

Plackett-Burman experiment design (PBD) showed a wide variation in xylitol production. This disparity in xylitol production reflected the importance of optimization to attain higher productivity. From the Pareto chart, (Figure 9.1) the variables, viz. xylose, potassium dihydrogen phosphate, yeast extract and magnesium sulphate were selected based on the 'frequency of occurrence' principle shown above the line of significance level. Further, the level of factors xylose, potassium dihydrogen phosphate, yeast extract and magnesium sulphate and the effect of their interactions on xylitol production were determined by face-centred central composite design of RSM.

Table 9.2 Face-centred central composite design (FC-CCD) experiment and response.

Run	A Xylose	B Yeast extract	C KH$_2$.PO$_4$	D MgSO$_4$	Response Xylitol yield (g/g) Experiment	Predicted
1	100	4	3	2	0.55	0.54
2	100	4	3	2	0.58	0.57
3	100	2	3	2	0.39	0.38
4	100	4	3	3	0.45	0.44
5	100	6	3	2	0.45	0.44
6	120	2	2	1	0.19	0.19
7	100	4	3	2	0.57	0.56
8	120	2	4	3	0.15	0.15
9	100	4	3	2	0.56	0.55
10	80	6	4	3	0.15	0.15
11	100	4	3	1	0.42	0.41
12	100	4	4	2	0.46	0.45
13	80	6	4	1	0.07	0.07
14	80	2	2	3	0.09	0.09
15	100	4	3	2	0.58	0.57
16	120	2	2	3	0.23	0.23
17	80	2	2	1	0.08	0.08
18	100	4	3	2	0.58	0.57
19	80	4	3	2	0.45	0.44
20	120	6	4	1	0.12	0.12
21	80	2	4	1	0.09	0.09
22	80	2	4	3	0.19	0.19
23	120	6	4	3	0.17	0.17
24	80	6	2	3	0.14	0.14
25	80	6	2	1	0.09	0.09
26	120	4	3	2	0.49	0.48
27	120	6	2	1	0.26	0.25
28	100	4	2	2	0.48	0.47
29	120	6	2	3	0.26	0.25
30	120	2	4	1	0.08	0.08

Thirty experiments were performed at different combinations (within the concentration range as derived from PBD results) of the factors shown (Table 9.2) and the central point was repeated six times (1, 2, 7, 9, 15 and 18) in triplicates to reduce standard error. The predicted and observed responses along with design matrix are presented in Table 9.2. The results were analyzed using Analysis of Variance (ANOVA). The second order regression equation provided the levels of xylitol production as a function of xylose, potassium dihydrogen phosphate, yeast extract and magnesium sulphate, which can be presented in terms of coded factors as mentioned in the following equation:

$$Y = 0.56 + 0.03*A + 0.01*B - 0.018*C + 0.023*D + 0.01*AB - 0.032*AC$$
$$-0.005*AD - 0.01*BC - 0.0025*BD + 0.0125*CD - 0.082*A^2$$
$$-0.13*B^2 - 0.082*C^2 - 0.117*D^2 \tag{9.3}$$

where Y is the xylitol yield in weight basis (g/g), A, B, C and D are xylose, potassium dihydrogen phosphate, yeast extract and magnesium sulphate, expressed in w/v basis, respectively.

Box Benhem design was used for optimization of the process variables namely agitation (rpm), Temperature (°C), pH and Inoculum level (ml/l) for increased xylitol yield. Ranges of the four process variables assessed at 4 coded levels as denoted in Table 9.3. The xylitol yield was selected as response due to different cycle of runs. Totally twenty nine experiments were performed in triplicates to analyze the response abd minimizes standard error.

The quadratic models in terms of coded variables are shown in the following equation (4), where (Y) represents xylitol yield (g/g), as a function of agitation (A), pH (B), Temperature (C) and Inoculum level (D).

$$Y = 0.534 + 0.04*A + 0.013*B - 0.02*C + 0.017*D + 0.0025*AC$$
$$- 0.02*AD + 0.017*BC + 0.037*BD + 0.04*CD - 0.137*A^2$$
$$- 0.193*B^2 - 0.173*C^2 - 0.142*D^2 \tag{9.4}$$

To fit the response function and experimental data, regression analysis was performed and the second order model for the response was assessed by ANOVA. The regression for the response was statistically substantial. For the response (Y), the model did not show any lack of fit and determination coefficient (R^2) for xylitol production obtained was 0.9954, which explained 99% of the variability in response. The predicted R^2 value of 0.9787 was in rational agreement with the adjusted R^2 value of 0.9912. An adequate precision value greater than 4 is desirable. The adequate precision value of 39.116 indicates an adequate signal and suggests that the model can be to navigate the design space. There is only a 0.01% chance that a "Model F-value" this large could occur due to noise. The smaller the magnitude of the 'P', more significant is the corresponding coefficient. 'P' value less than 0.05 indicate the model terms are vital. From the 'P' value it was found that, the variables, A, B, C, D, AB, AC, BC, CD, A^2, B^2, C^2, D^2 were significant for xylitol production. The above model can be used to predict

Table 9.3 Box Behnkem design experiment and response.

Run	A Agitation (rpm)	B pH	C Temperature (°C)	D Inoculum level (ml)	Response Xylitol yield (g/g) Experiment	Predicted
1	100	4	30	0.5	0.21	0.20
2	150	5	30	1.5	0.3	0.28
3	50	4	30	1	0.16	0.15
4	150	6	30	1	0.25	0.24
5	100	6	30	1.5	0.25	0.24
6	100	6	30	0.5	0.16	0.15
7	100	5	30	1	0.52	0.49
8	100	5	28	1.5	0.24	0.23
9	50	5	30	0.5	0.18	0.17
10	150	5	32	1	0.23	0.22
11	100	6	32	1	0.19	0.18
12	100	6	28	1	0.19	0.18
13	100	5	32	1.5	0.28	0.26
14	100	5	30	1	0.53	0.50
15	100	5	30	1	0.54	0.51
16	50	5	32	1	0.16	0.15
17	100	4	32	1	0.12	0.11
18	50	5	30	1.5	0.23	0.22
19	100	4	30	1.5	0.15	0.14
20	50	6	30	1	0.18	0.17
21	100	5	30	1	0.55	0.52
22	100	5	30	1	0.53	0.50
23	150	5	30	0.5	0.33	0.31
24	100	5	32	0.5	0.12	0.11
25	100	4	28	1	0.19	0.18
26	150	5	28	1	0.27	0.25
27	150	4	30	1	0.23	0.22
28	50	5	28	1	0.21	0.20
29	100	5	28	0.5	0.24	0.23

the xylitol production within the limits of the experimental factors that the actual response values agree well with the predicted response values.

The above model is used to predict the xylitol production within the limits of the experimental factors that the actual response values agree well with the predicted response values. The production of xylitol with variable inter-actions were studied by plotting surface curves against any two independent variables, while keeping another variable at its central (0) level. The curves of

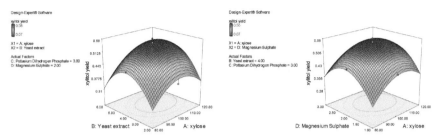

Figure 9.2　Three dimensional plot showing interactive effect of yeast extract and xylose and magnesium sulphate and xylose on xylitol yield.

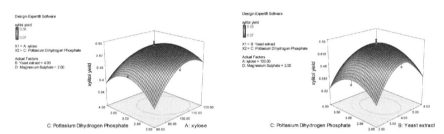

Figure 9.3　Three-dimensional plot showing interactive effect of KH_2PO_4 and xylose and KH_2PO_4 and yeast extract on xylitol yield.

calculated response (xylitol production) and contour plots from the interactions between the variables are shown in Figures 9.2–9.4. Figure 9.2 show the dependency of xylitol on xylose against yeast extract and magnesium sulphate. The xylitol production increased with increasing in yeast extract to about 4 g/l and then xylitol production is decreased with further increase in yeast extract. Similar results were observed in Figures 9.3 and 9.4. Increase in KH_2PO_4 ensued a prominent rise in xylitol production upto 2.8 g/l. The optimal conditions of $(NH_4)_2SO_4$, KH_2PO_4, $MgSO_4 \cdot 7H_2O$ and yeast extract for maximum xylitol production were determined by contour plot analysis and estimated by regression equation.

9.2.4　Validation of the Experimental Model

It was tested by carrying out the batch experiment under optimal operation conditions: Xylose $\sim$104 g/l, Yeast Extract $\sim$4.0 g/l; KH_2PO_4 $\sim$2.8 g/l and $MgSO_4$ $\sim$2. g/l established by the regression model. Four repeated experiments were performed and the results are compared. The xylitol production (0.555g/g) obtained from experiments was very close to the actual response

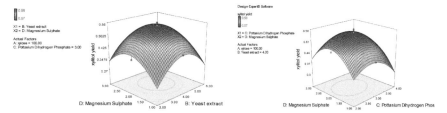

Figure 9.4 Three-dimensional plot showing interactive effect of magnesium sulphate and yeast extract and KH_2PO_4 and magnesium sulphate on xylitol yield.

(0.568 g/g) predicted by the regression model, which proved the validity of the model.

To fit the response function and experimental data, regression analysis was performed and the second order model for the response was evaluated by ANOVA. The regression for the response was statistically significant. For the response (Y), the model did not show any lack of fit and determination coefficient (R^2) for xylitol production obtained was 0.9890, which explained 98% of the variability in response. The predicted R^2 value of 0.9411 was in sensible agreement with the adjusted R^2 value of 0.9780. An adequate precision value greater than 4 is desirable. The adequate precision value of 29.334 indicates an adequate signal and suggests that the model can be to navigate the design space. There is only a 0.01% chance that a "Model F-value" this large could occur due to noise. Also the model F value of 90.09 implies the model is noteworthy. The smaller the magnitude of the 'P', more significant is the corresponding coefficient. 'P' value less than 0.05 indicate the model terms are significant. From the 'P' value it was found that, the variables, A, B, C, D, BD, CD, A^2, B^2, C^2, D^2 were significant for xylitol production. The above model can be used to predict the xylitol production within the limits of the experimental factors that the actual response values agree well with the predicted response values.

The response surface curves for the xylitol yield were shown in Figures 9.5–9.7. The nature of the response surface curves illustrates the interaction between the variables. From the diagrams, it was observed that increase in the agitation beyond 107 rpm reduces the xylitol yield. Similar patterns were detected for the Figures 9.6 and 9.7. In general, the elliptical shape of the curve indicates decent interaction between the two variables and circular shape indicates no interaction between the variables. From the figure, it is observed that elliptical nature of the contour in graphs portrays the mutual interaction of all the variables. There is a relative significant

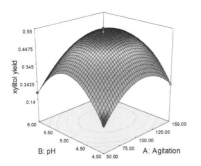 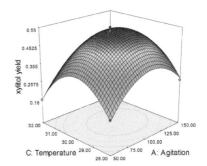

Figure 9.5   Three-dimensional plot showing interactive effect of agitation and pH and agitation and temperature.

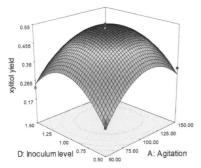 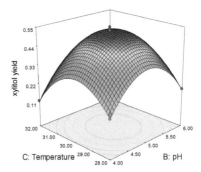

Figure 9.6   Three-dimensional plot showing interactive effect of agitation and inoculum level and temperature and pH.

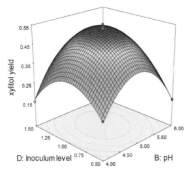 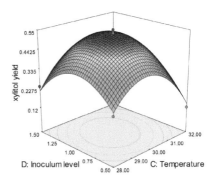

Figure 9.7   Three-dimensional plot showing interactive effect of inoculum level and pH and inoculum level and temperature.

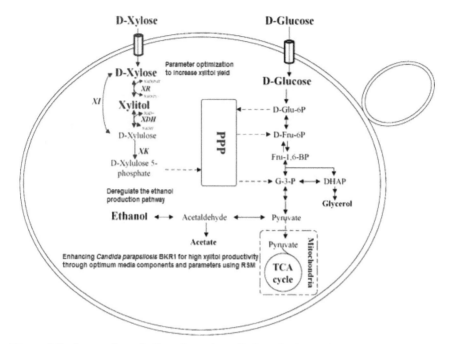

Figure 9.8   Proposed metabolic pathway for xylitol production and the effects of optimiza-tion inside *Candida parapsilosis* strain BKR1.

interaction between every two variables, and there is a maximum predicted yield as designated by the surface limited in the smaller ellipse in the contour diagrams.

Validation of the experimental model was confirmed by carrying out the batch experiment under optimal operating conditions, Agitation: 107 rpm, pH: 5, Temperature: 30°C, Inoculum level: 1 ml established by the BBM resulted regression model. Two repeated experiments were performed and the results are compared. The xylitol production (0.527 g/g) obtained from experiments was very marginal to the actual response (0.538 g/g) predicted by the regression model, which proved the validity of the model.

9.3 Conclusion

In this work, Plackett-Burman design was used to determine the relative importance of medium components for xylitol production. Among the vari-ables, Potassium Dihydrogen Phosphate, Magnesium sulphate, yeast extract and xylose were found the most significant variables. From further optimization studies the optimized values of the variables for xylitol production were as follows: Xylose ~104 g/l, Yeast Extract ~4 g/l; KH_2PO_4 ~2.8 g/l and $MgSO_4$ ~2 g/l. This study showed that significance of medium components during the mass production of xylitol. Besides the operation conditions, Agitation: 107 RPM, pH – 5, Temperature – 29.9°C, Inoculum level – 1 ml were optimized. Based on the optimized conditions, the production reaches 0.55 g/g. The results show a close agreement between the expected and obtained production level.

The multiphase stirred tank reactor developed in this study has the ability to produce 0.55 g/g yield of xylitol against xylose in the indigenously isolated *Candida parapsilosis* strain BKR1. Intermittently the productivity has to be improvised to meet the industrial feasibility. Indisputably the research findings reveal that *Candida species* are best xylitol producers by consuming xylose as substrate. Figure 9.8 shows the probable metabolic pathway followed by the *Candida parapsilosis* strain BKR1 to produce xylitol. To improve productivity of xylitol, the ethanol production pathway has to be inhibited by targeting suitable enzyme inhibitors and also the rate of xylose reductase to be enhanced for high xylitol productivity. Optimization of the medium components and process parameter plays a vital role in fermentative production of xylitol, a natural polyol and important pharmaceutical constituent.

Acknowledgements

Authors would like to express their thanks to DST, DBT – GoI, Management of Bannari Amman Institute of Technology for their financial support and providing the laboratory facilities.

References

[1] A. Bar, LO. Nabors, RC. Gelardi, 'Xylitol in Alternative sweetener', Marcel Dekker Inc., pp. 349-379, 1991.
[2] A. Emodi, 'Xylitol: Its Properties and Food Application', Food Technol., pp. 20-32, 1978.
[3] DC. Montgomery, 'Design and Analysis of Experiments', John Wiley and Sons, New York., pp. 124-127, 2001.

[4] E. Winkelhausan, S. Kusmanova, 'Microbial conversion of D-xylose to xylitol', J. Ferment. Bioeng, pp. 1-14, 1998.

[5] F. Chiung, Huang Yi-Feng Jiang, Gia- Luen Guo, 'Method for producing xylitol from Lignocellulosic hydrolysates without Detoxification', United states patent and publication., 2014.

[6] Jain, T., Grover, K. Sweeteners in Human Nutrition. Int. J. Health Sci. Res., 5(5): 439-51. 2015.

[7] G. Koutitas, P. Demestichas, 'A review of energy efficiency in telecommunication networks', Proc. In Telecomm. Forum (TELFOR)., pp. 1-4, Serbia, Nov., 2009.

[8] I. Cerutti, L. Valcarenghi, P. Castoldi, 'Designing power-efficient WDM ring networks', ICST Int. Conf. on Networks for Grid Applic., Athens, 2009.

[9] Arora, R., Behera, S., Sharma, N.K., Kumar, S. A new search for thermotolerant yeasts, its characterization and optimization using response surface methodology for ethanol production. *Front Microbiol.*, 2015. http://dx.doi.org/10.3389/fmicb.2015.00889.

[10] Ghindea, R., Csutak, O., Stoica, I., Ana Maria Tanase, Vassu, T. Production of xylitol by yeasts. Rom Biotech Let., 15: 3, 2010.

[11] J. Haas, T. Pierce, E. Schutter, 'Datacenter design guide', Whitepaper, the Greengrid, 2009.

[12] JM. Dommguez, CS. Gong, G. Tsao, 'Production of xylitol from D-xylose by *Debaryomyces hansenii*', Appl. Biochem. Biotechnol, pp.117-127, 1997.

[13] L. Hyvonen, P. Slotte, 'Properties and Food Application', Food Technol., pp. 97-112, 1983.

[14] MY. Noordin, VC. Venkatesh, S. Sharif, S. Elting, A. Abdullah, 'Application of response surface methodology in describing the performance of coated carbide tools when turning AISI 104 steel', J. Mater. Process. Technol., pp.46-58, 2004.

[15] Kavsceek, M., Strazar, M., Curk, T., Natter, KK., Petrovic, U. Yeast as a cell factory: current state and perspectives. Microb. Cell Fact., 14(94): 1-10. 2015.

[16] P. Converti, A. Perego, Sordi, P. Torre, 'Effect of starting xylose concentration on the microaerobic metabolism of *Debaromyces hanseneii*: the use of carbon material balance', Biochem. Biotechnol., pp.15-29, 2002.

[17] R. Balakrishnaraja, P. Ramalingam, K. Selvapriya, 'Molecular identification and single factorial optimization of microbial isolates for natural polyol production', Rasayan J. Chem., pp. 241-245, 2014.

[18] RS. Rao, RS. Prakashan, K. Krishna Prasad, S. Rajesham, PN. Sarma, LV. Rao, 'Xylitol production by *Candida* sp.: parameter optimization using Taguchi approach', Process Biochem., pp. 951-956, 2004.

[19] Kogje, A., Ghosalka, A. Xylitol production by *Saccharomyces cerevisiae* overexpressing different xylose reductases using non-detoxified hemicellulosic hydrolysate of corncob. *3 Biotech.*, 6: 127. 2016.

[20] T. Walther, P. Hensirisak, FA. Agblevor, 'The influence of aeration and hemicellulosic sugars on xylitol production by *Candida tropicalis*', Bioresour. Technol., pp. 213-220, 2001.

[21] W. Li, W. Du, DHJ. Liu, 'Optimization of whole cell catalysed methanolysis of soybean oil for biodiesel production using response surface methodology' Mol. Catal., pp. 122-127, 2007.

[22] W. Vereecken, L. Deboosora, P. Simoens, B. Vermeulen, D. Colle, C. Develder, M. Pickavet, B. Dhoedt, P. Demeester, 'Energy Efficiency in thin client solutions', ICST Int. Conf. on Networks for Grid Applic., Athens, 2009.

[23] Y. Takahashi, C. Takeda, I. Seto, G. Kawano, Y. Machida, 'Formulation and evaluation of lactoferrin bioadhesive tablets', Int. J. Pharmacol., pp. 220-227, 2007.

Part IV

Nanostructured Polymers for Biomedical Applications

10

Self-assembled Nanostructures of Polysaccharides for Therapeutics

V. S. Prasad[*], **Aarsha Surendren, P. Anju, Asha Susan Chacko**
and Sumesh Soman

Materials Science and Technology Division, Council of Scientific and
Industrial Research – National Institute for Interdisciplinary Science and
Technology, Thiruvananthapuram, India
E-mail: vsprasad@niist.res.in

Nanomaterials of natural biopolymers such as polysaccharides are receiving increased attention in controlled therapeutics. The possibility of regioselective functionalisation and manipulation of ionisability makes the polysaccharide a unique candidate for targeted and stimuli-responsive delivery with biocompatibility and non-toxic nature, especially for cancer chemotherapy. Self-assembly of the suitably modified amphiphilic polysaccharide homopolymers and copolymers could be used as a facile synthetic strategy for nanostructures such as micelles, vesicles or tubes for effective drug targeting and as an advanced treatment methodology. Modified nanocellulose-based self-assembled vesicles having tailor-made size and wall thickness by controlled long alkenyl modification with efficient drug loading capability and stimuli-responsive delivery can be used as biocompatible and non-toxic nanostructures in chemotherapy. The limitations in traditional therapeutic and diagnostic agents can be solved by the unique physicochemical properties of these nanomaterials including controlled size in nano-/microdimensions, large surface area to mass ratio coupled with high reactivity compared to bulk materials. The polysaccharide-based nanostructures can address the toxicity concerns of the bulk materials and metal nanoparticles in therapeutics.

193

10.1 Introduction

Nanomaterials are finding applications in many advanced areas, especially in therapeutic applications. In this case metallic nanoparticle assisted directed drug delivery is of great potential. The storage and delivery of drugs in polymeric micelles or self-assembled vesicles are of recent scientific interest owing to control of the vesicle size and storage capacity in addition to the possibility of control in drug delivery [56]. Self-assembly is a spontaneous process by which organised structures with particular functions and properties could be obtained without additional cumbersome processing or modification steps [32]. Polysaccharide-based nanostructures such as tubes, vesicles, micelles, and other self-assembled morphologies are of great value in biomedical applications owing to the non-toxic nature and biocompatibility.

Nanotechnology is the understanding and control of matter generally in the 1–100nm dimension range. In nanomedicine where the strategies of nanotechnology are applied to medicine, concerns with the use of precisely engineered materials at this length scale to develop novel therapeutic and diagnostic methodologies [75]. The limitations in traditional therapeutic and diagnostic agents can be solved to some extent by the unique physicochemical properties of these nanomaterials, such as ultra-small size, large surface area to mass ratio coupled with high reactivity compared to bulk materials (Figures 10.1 and 10.2).

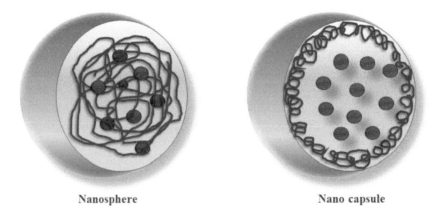

Nanosphere   Nano capsule

Figure 10.1   Typical polymeric nanoparticles.

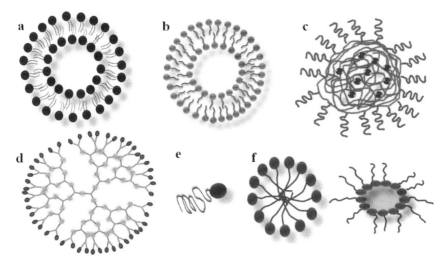

Figure 10.2   Schematic illustration of Nanoparticle platforms for therapeutics, (a) liposome, (b) niosome, (c) polymeric nanoparticle (d) dendrimer (e) polymer–drug conjugate (f) micelles.

Self-assembly of amphiphilic block co-polymers is reported to form different morphologies in nano or micro dimensions including spheres, vesicles, and tubes. The copolymers with tailor-made amphiphilic properties or natural polymers with regio-selective aliphatic modifications can be utilised for biomedical applications. Among these, self-assembled polymeric micelles have been the subject of many studies in the field of drug delivery [4]. Polymeric micelles of block copolymers of poly (ethylene oxide) with poly (amino acids) have shown great potential in drug delivery, drug solubilisation and in targeted delivery (Figure 10.3). Kim et al. presented the integration of amphiphilic block copolymer micelles of poly (acrylic acid) (PAA) and poly (ethylene oxide)-block-poly(-caprolactone) (PEO-b-PCL) as nanometre-sized vehicles for hydrophobic drugs within layer-by-layer films using alternating hydrogen bond interactions as the driving force for assembly, enabling the incorporation of drugs for pH-dependent release [29].

Since direct injection of drugs may cause side effects due to their permeation to other regions of the body, concealment and targeting with appropriate materials is a critical consideration in the design of practical drug delivery systems. Stimuli-free auto-modulated drug release using layered assembly of micro and nanoshells was experimented by Ariga et al. [5]. The layer by layer

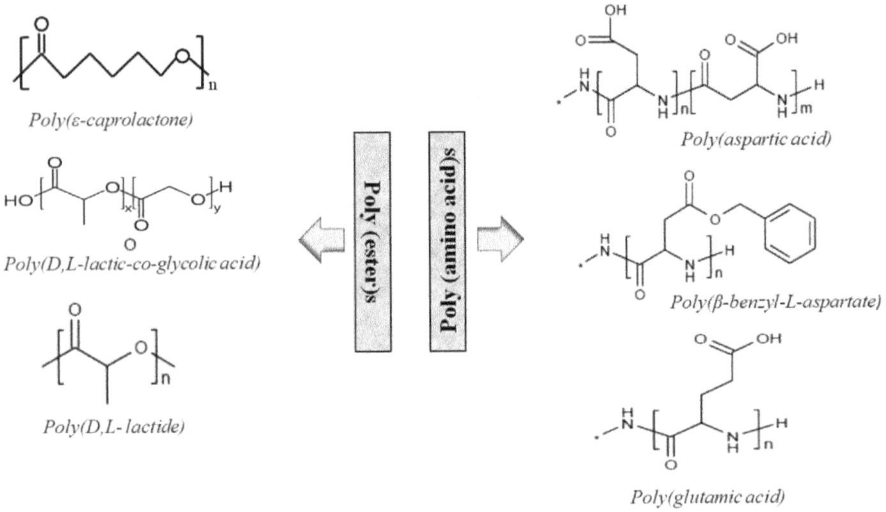

Figure 10.3   Polymers and Block copolymers used in drug delivery applications.

micro shell assembly of gelatin was utilised for drug delivery by Shutva et al. [60]. In these aqueous suspensions of insoluble drugs have been subjected to powerful ultrasonic treatment followed by sequential addition of polycations and polyanions to the particle solution leading to assembly of ultra-thin polyelectrolyte shells on the nano-sized drug particles. The stepwise release of drugs from LbL films of mesoporous capsules to the exterior in the absence of external stimuli was also demonstrated. Oral drug delivery through nanoparticles such as polymeric micelles is an important strategy for gastrointestinal drug administration due to the increased solubility of the drugs in polymer micelles [20].

Nanoparticles offer many advantages for drug and gene delivery as well as medical diagnostics and imaging [34]. The size of the polymer particle-drug conjugate plays an important role in determining the bioavailability and in turn the efficacy of the drug. Also, they are large enough to avoid glomerular permeation and small enough to avoid clearance via uptake by macrophages. The particle diameters should be <500 nm to escape phagocytosis [15]. The small size allows the particles to pass through the tissue gap and capillaries by diffusion. Nanoparticles are also more resistant to hepatic filtration than larger particles. These unique properties of polymeric nanoparticles have been utilised to control the release

Table 10.1 Self-assembled organic nanosystems that are currently used in clinical practice.

Platform type	Trade name	Formulation	Indication	Reference
Liposomes	Liposomes DaunoXome	Liposomal daunorubicin	• Breast cancer • Advanced HIV-associated Kaposi sarcoma • Acute nonlymphocytic leukemia • Ovarian cancer • multiple myeloma	[53]
	Marqibo	Liposomal vincristine	• Leukemia • Hodgkin diseas • Non-Hodgkin lymphomas • Wilms'tumor • Rhabdomyosarcoma • Neuroblastoma	[17]
	Halaven	Liposomal eribulin mesylate	• Breast cancer • Liposarcoma	[68]
Micelles	Genexol-PM	PEGylated micellar paclitaxel	• Breast canter • small cell lung cancer	[9]
Supramolecular protein-drug formulation	Abraxane	Albumin-bound paclitaxel	• Breast cancer • Pancreatic cancer • Nonsmall-cell lung cancer	[62]

of macromolecular drugs, proteins, and vaccines and in controlled-release applications.

Self-assembled hybrid nanostructures as nanoplatforms for cancer diagnosis and therapy was reviewed recently [33]. Self-assembled organic-inorganic hybrids are a class of promising nanomaterials possessing interesting physiochemical and biological characteristics, making them highly attractive in cancer-related biomedical applications. Using various organic-inorganic species as the building blocks one can create a myriad of hybrid nanosystems using self-assembly strategy which will show favourable interfacial properties retaining their physical or chemical functionalities (**Table 10.1**).

Advances in the design of nanoscale stimuli-responsive systems that are able to control drug biodistribution in response to specific exogenous stimuli such as temperature, magnetic field, ultrasound intensity, light or electric pulses or endogenous variants including changes in pH, enzyme concentration or redox systems were reviewed by Mura et al. [47]. Stimuli-responsive drug delivery systems that deliver a drug in spatial-, temporal- and dosage-controlled fashions have become the progressing field in therapeutics. Implementation of such devices requires the use of biocompatible materials that are susceptible to a specific physical incitement or that, in response to a specific stimulus, undergo a protonation, a hydrolytic cleavage or a supramolecular conformational change. Polysaccharides in nanoforms which are modifiable by functionalisation are the best candidates for applications in stimuli-responsive drug delivery [6].

10.2 Polysaccharides for Drug Delivery

Polysaccharides are gaining increasing attention as components of stimuli-responsive drug delivery systems since they can be obtained from natural renewable sources [41]. Polysaccharides-based nanoparticles, are produced by covalent crosslinking, ionic crosslinking, polyelectrolyte complex formation and the self-assembly of their hydrophobically modified systems [40]. They are promising alternatives to polyethylene glycol (PEG) which is widely used in biomedical applications because they contain a variety of functional groups (hydroxyl, amino, and carboxylic acid) that can be used for drug conjugation and self-assembly [48]. Many polysaccharides are polyelectrolytes also such that the surface charge of polysaccharide carriers can be used to engineer bio-interactions such as cellular uptake or glomerular filtration [12]. Cationic polysaccharides promote endocytic uptake by cells, whereas anionic polysaccharides could increase bioavailability by reducing excretion through the glomerular capillary wall. Negatively charged species would be less filterable because of the negative charge of the glomerular membrane. Control of both these functionalities could be used for applications as charge-reversible and pH-responsive polysaccharides.

Ionic polysaccharides can be made into cross-linkable hydrogels sensitive to external stimuli to control the drug release pattern. The applications can be further extended by making inorganic-organic hybrids or composites or by suitable grafting of functional groups to reinforce the responsive character. Polysaccharide based aerogels result in highly porous (90–99%) and light weight (0.07–0.46 g/cm^3) drug carriers with large surface area (Sa = 70–680 m^2/g) for enhanced drug loading capacity resulting in increased bioavailability of the drug [19].

10.3 Self-assembled Nanostructures

10.3.1 Micelles

Self-assembled micelles are investigated to improve drug solubility, and stability since the delivery of hydrophobic molecules and proteins has been an issue due to poor bioavailability by Zhang et al. [74]. Natural polysaccharides are being explored as substitutes for synthetic polymers in the development of new micelle systems due to problems with toxicity and immunogenicity. Self-assembled micelles can be readily formed in aqueous solution by grafting hydrophobic moieties to the polysaccharide backbone [25]. Many polysaccharides also possess inherent bioactivity that can facilitate mucoadhesion, enhanced targeting of specific tissues, and a reduction in the inflammatory response. The hydrophilic nature of polysaccharides can be utilised to enhance circulatory stability.

Among the polymeric micelles, hydrophobised polysaccharides have currently become one of the hottest researches in the field of drug delivery nanosystems. The characteristic properties of these nanosystems such as small particle size with narrow size distribution, specific core-shell structure, high solubilisation, structural stability, passive tumour localisation by enhanced permeability and retention (EPR) effect, active targeting ability via tailored targeting groups, in addition to the facile synthetic strategy. The polymeric micelles self-assembled by hydrophobised polysaccharides by stereo and regio-specific modification can be employed as targeted drug delivery nanosystems by including thermo- or pH-sensitive components or by attaching specifically targeted moieties to the outer hydrophilic surface [39]. Hydrophobised polysaccharide polymeric micelles can also serve as an effective non-viral vector for gene delivery, besides encapsulation of water-insoluble drugs, which can complex with charged proteins or peptide drugs through electrostatic force or hydrogen bond.

10.3.2 Vesicles

Supramolecular chemistry has been developed for the synthesis and self-assembly of natural materials into hollow spheres called vesicles. The vesicles are at the same time one of the main nano/micro structures in living organisms. The self-assembly of amphiphilic biomolecules based on lipids, proteins, and carbohydrates leads to this type of nanostructures. Large vesicles act as membranes to protect the intracellular components from the extracellular environment, whereas the small vesicles act in the intracellular transport of biomolecules. Stable vesicular structures have been reproduced in the laboratory, as early as the 1960s with liposomes assembled from naturally occurring phospholipids, and more recently with so-called polymersomes, which result from the self-assembly of block copolymers into vesicles [10]. Polymersomes are expected to play a tremendous role in the development of biomedical applications, such as the delivery of therapeutics, and as microreactors that mimic the behaviour of living cells.

Polymersomes hold some advantages over liposomes, such as a lower critical aggregation concentration, a higher membrane viscosity, and elasticity. These characteristics enhance membrane stability and decrease the passive diffusion of encapsulated molecules. A greater chemical diversity for surface functionalisation is also expected. Most of the polymersomes were assembled from synthetic amphiphiles with biocompatible or bioresorbable blocks, such as polylactide-block-poly (ethylene oxide), polycaprolactone-block-poly (ethylene oxide), and poly(2-methyl oxazoline)-block-poly(dimethyl siloxane)-block poly(2-methyl oxazoline).

The trend in drug delivery is consequently directed towards integrated multifunctional carrier systems, providing selective recognition in combination with the sustained or triggered release to achieve the best therapeutic effects, where it should carry the optimum amount of a drug to the desired target where it should be released at the optimum rate for a specified time. Vesicular systems can effectively confine the drugs for controlled release. Furthermore, carriers modified with recognition groups as tags can enhance the capability of drug delivery for a better therapeutic effect. Cui et al. reviewed the advances regarding designing and preparing assembled vesicles with controllable size with targeting ligands for selective recognition in drug delivery [15].

A glycoprotein analogue in which a polysaccharide block of dextran is linked linearly to a polypeptide block of poly (γ-benzyl L-glutamate) was also reported by Schatz et al. demonstrating that these macromolecular structures can self-assemble spontaneously into vesicles in a water medium [59]. This showed the applicability in the field of drug and gene delivery applied to a large range of polypeptide and polysaccharide molecules of biological interest, such as poly(sialic acid) or hyaluronic acid. The ability of these copolymers to self-assemble into small vesicles could be used to construct a new generation of drug and gene-delivery systems with a high affinity for the surface glycoproteins of living cells. This type of capsular structure composed of molecules of a polysaccharide-block-polypeptide can lead to systems that mimic virus morphology.

Colon-specific drug delivery was also reported using natural polysaccharides which were extensively used for the development of solid dosage forms for delivery of drug to the colon since the presence of large amounts of polysaccharides in the human colon as the colon is inhabited by a large number and variety of bacteria secreting many enzymes such as D-glucosidase, D-galactosidase, amylase, pectinase, xylanase, D-xylosidase, dextranase, etc. [61, 69, 70].

10.3.3 Giant Vesicles

Self-assembly of macromolecules is fundamental to life itself, and historically, these systems have been primitively mimicked by the development of amphiphilic systems, driven by the hydrophobic effect. Brosnan et al. demonstrated that self-assembly of purely hydrophilic systems into giant vesicles for drug delivery could also be achieved with polysaccharides such as pullulan (**Brosnan et al. 2015**). They have synthesised double hydrophilic block copolymers from polysaccharides and poly (ethylene oxide) or poly(sarcosine) to yield high molar mass diblock copolymers through oxime chemistry. These hydrophilic materials easily assembled into nanosized (<500 nm) and micro-sized (>5 μm) polymeric vesicles depending on concentration and diblock composition. Solely hydrophilic nature of these materials with extraordinarily water permeability showed prospects for applications as cellular mimics.

Many polysaccharides including dextran, chitosan, alginates, cellulose, etc. were tried as nano drug conjugates in drug delivery applications as shown in Table 10.2 [31]. A detailed classification of such polysaccharides is shown in Figure 10.4. Dextrans for targeted and sustained delivery of therapeutic and imaging agents were reported [43].

10.4 Other Polysaccharides Used in Drug Delivery

10.4.1 Dextrin

Dextrin is a low molecular weight polysaccharide obtained from starch by hydrolysis using enzymes such as amylases [24]. Sridhar et al. reported for the first time on the self-assembled vesicles of dextrin and their applications in drug delivery (Figure 10.6, [67]). Cyclodextrin-based multiwalled capsules for drug delivery were reported by Qi et al. [57].

Table 10.2   Polysaccharide-based nanomaterials for target-based drug delivery applications.

Nanomaterials	Synthesis Methods	Drug	Application	References
Alginate				
Alginate gel beads-entrapped liposome	Encapsulation	Bee venom peptide	n.s.	Xing et al. [72]
Chitosan				
Chitosan microspheres	Spray-drying	Carbamazepine	Nasal delivery	Gavini et al. [21]
Chitosan (CS) and poly(gamma-glutamic acid) nanoparticles	Ionotropic gelatin	Insulin	Diabetes therapy	Lin et al. [37]
Hyaluronic acid-coupled chitosan nanparticles	Ionotropic gelation	Oxaliplatin	Anticancer therapy	Jain et al. [27]
N-trimethyl chitosan-hyaluronan-cisplatin conjugate nanoparticles	Ionotropic gelation	Cisplatin	Anticancer therapy	Cafaggi et al. [11]
Chitosan nanoconstructs	Ionotropic gelation	Heparin	Oral delivery	Paliwal et al. [41]
Dextran				
Dextran magnetite and conjugated cisplatin	Thermal ablation	Cisplatin	Anticancer therapy	Sonoda et al. [66]
Folic acid-dextran-camptothecin nanoparticles	Supercritical antisolvent	Camptothecin	Anticancer therapy	Zu et al. [76]
Gellan gum and Xanthan gum				
Gellan gum-PVA hydrogel microspheres	Encapsulation	Carvedilol	Antihypertensive, Oral drug delivery	Agnihotri and Aminabhavi [3]
Gellan gum beads	Ionotropic gelation	Cephalexin	Oral delivery	Agnihotri et al. [2]

10.4.2 Chitosan

Chitosan is obtained by the controlled deacetylation of chitin, an abundant amino polysaccharide obtained from the crustacean shells. Polymer-drug conjugates of chitosan-based nanomaterials have attained recent attention as due to good biocompatibility, biodegradability, and low toxicity. These systems have been prepared by emulsion techniques or by chemical or ionic gelation, coacervation/precipitation, and spray-drying methods. As

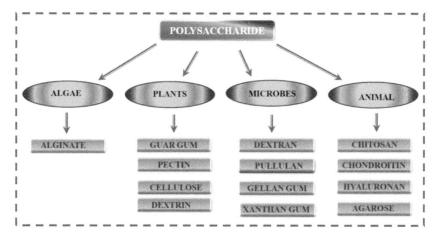

Figure 10.4 Classification of polysaccharides.

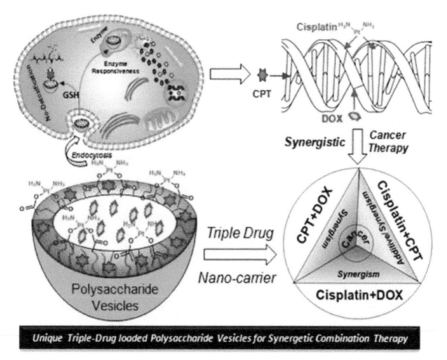

Figure 10.5 Novel cisplatin-stitched polysaccharide vesicular nanocarrier design for the synergistic combination therapy of antagonistic drugs [16].

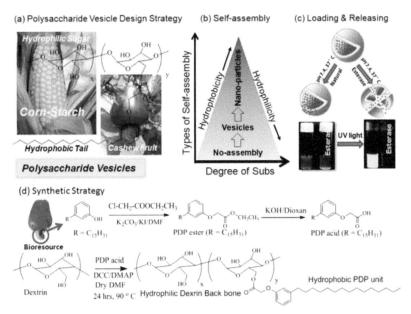

Figure 10.6   (a) Dextrin vesicle design strategy, (b) self-assembly of dextrin with respect to hydrophilic and hydrophobic content, (c) loading and releasing capabilities of dextrin vesicles under normal and esterase enzyme and (d) synthetic methodologies [67].

alternatives to the traditional fabrication methods, self-assembled chitosan nanomaterials show significant advantages [73].

10.4.3 Alginates

Alginates are linear unbranched natural polysaccharides extracted from brown seaweeds and marine algae. It is an anionic copolymer comprising of (1→4) linked β-D-mannuronic acid and α-L-guluronic acid units of different composition and sequence depending on the alginate source. Alginate shows variable molecular weight, and degree of polymerisation dependent on the enzymatic control during the production. Commercial alginates exhibit a typical average molecular weight of approximately 200,000 Da. [58]. Alginate can be designed as a controlled-release product based on its favourable pH-sensitive properties for drug delivery. Polymer-coated alginate microspheres and nanoparticles have proved promising for modified drug delivery systems giving mechanical stability and the drug release control [42]. Different drugs, such as indomethacin [54], prednisolone [71], metoclopramide hydrochloride, antiemetic drug [21, 22], protein drugs [23, 35], diclofenac [50], antidiabetic drug, probucol [44], resveratrol [28] and risperidone [8] have been investigated as drug delivery agents for prolonged and better drug release at the target site.

The drug-loaded nanospheres and vesicles were prepared by self-assembly of alginate in aqueous media containing Ca^{2+} and CO_3^- ions under very mild conditions by adjusting the preparation conditions, nano-sized drug-delivery systems such as nanospheres and vesicles were obtained. The morphologies of the drug-delivery systems with the anticancer drug, 5-fluorouracil, encapsulated in the nanospheres and vesicles, and in vitro drug release behaviour was reported [73].

10.4.4 Gellan Gum

Gellan gum is a high molecular weight anionic linear polysaccharide pro-duced by fermentation from *Sphingomonas paucimobilis*. This polymer is a tetrasaccharide repeating unit of D-glucose, l-rhamnose, and D-glucuronate in a molar ratio of 2:1:1 [18]. Chemical hydrogels of gellan are prepared using chemical crosslinking of networks to increase their mechanical proper-ties and enable slower drug release profiles [13,14]. In view of its unique structure and beneficial properties, gellan is currently being used for the development of hydrogels for controlled release forms in different various pharmaceutical formulations including oral, ophthalmic, nasal, and other treatments [51].

10.4.5 Xanthan Gum

Xanthan gum is fermentation-derived natural, high molecular weight polysac-charide from *Xanthomonas campestris*. It consists of $(1 \rightarrow 4)$ linked β-D-glucose units, having a trisaccharide side chain attached to alternate D-glucosyl residues. Although the gum has many suitable properties desirable for drug delivery, it has limited practical use with respect to the unmodified forms due to slow dissolution and substantial swelling in biological fluids. Chemical modification of xanthan gum is achieved by different conventional chemical methods like carboxymethylation, and grafting, such as free radical, microwave-assisted, chemo-enzymatic, and plasma-assisted chemical graft-ing has resulted in the alteration of different physicochemical properties for diverse biological applications [7].

10.4.6 Pullulan

Pullulan is a natural polysaccharide, a maltotrios unit obtained from starch by the fungal *Aureobasidium pullulans*. Vitamin H (biotin) was incorporated into a hydrophobically modified polysaccharide, pullulan acetate (PA), in order to improve the cancer-targeting activity and internalisation of self-assembled nanoparticles. The biotinylated pullulan acetate (BPA) nanoparticles were prepared by a diafiltration method at 100 nm level and studied its loading efficiency by Naa et al. [49].

10.4.7 Cellulose

Self-assembled micelles based on hydrophobic modified quaternised water soluble cellulose for drug delivery was reported by Song et al. using cellulose obtained from cotton linter pulp [65]. The study on amphiphilic cationic cellulose (HMQC) derivatives carrying long chain alkyl groups as hydrophobic moieties and quaternary ammonium groups as hydrophilic moieties revealed that it can be self-assembled into cationic micelles in water with the average hydrodynamic radius of 320–430 nm. The cytotoxicity study showed that the HMQC exhibited low cytotoxicity. Prednisone acetate, a water insoluble anti-inflammation drug, was chosen as a model drug to investigate the utilisation of self-assembled HMQC micelles as a delivery carrier for poorly water-soluble drugs. The study indicated that the prednisone acetate could be incorporated effectively in the self-assembled HMQC micelles and showed prospects for use as a controlled release system.

10.5 Nanocellulose for Drug Delivery

Nanocellulose, a unique and promising natural material extracted from native cellulose, has gained much attention for its use as biomedical material, because of its remarkable physical properties, special surface chemistry and excellent biological properties such as biocompatibility, biodegradability and low toxicity. Cellulose nanocrystals (CNC), nanofibrillated cellulose (NFC) and bacterial cellulose (BC), the three different forms of nanocellulose for its production, properties and biomedical applications were reviewed [36]. CNC and NFC reported to bind and release water-soluble drugs via ionic interactions whereas BC has been used to release drugs from flexible membranes [55]. All forms of nanocellulose can be chemically modified to expand the range of drugs that may bind to the surface. Functional modification of nanocellulose will determine the potential biomedical application for nanocellulose.

Liu et reported a pH/near-infrared (NIR)-responsive hydrogel for on-demand drug delivery and wound healing based on nanocellulose [38]. Poly-dopamine (PDA) was introduced into CNF network to fabricate a PDA/CNF hydrogel through ion crosslinking with calcium ion as a cross-linker. Tetra-cycline hydrochloride (TH) which loaded on PDA could be released from the prepared hydrogels in an on-demand fashion under NIR exposure or at lower pH conditions for wound healing.

Potential applications of bacterial cellulose as transdermal drug delivery system were reviewed by Abeer et al. [1]. The properties of such a wound-healing system can easily be oriented towards transdermal drug delivery as it prevents moisture from evaporating, avoids external contamination and maintains intimate contact with the exposed, inflamed or diseased area. This facilitates localised drug delivery to the target site. Transdermal delivery systems that can work both to deliver drug as well as absorb exudates, gives an opportunity for the application of BC membranes.

Bacterial nanocellulose (BNC) membranes are used as the carrier for berberine hydrochloride and berberine sulphate to produce a new controlled release system [26]. The application of nanofibrillar cellulose as a matrix material for sustained drug delivery is reported [30]. BNC as drug delivery system for proteins using serum albumin as model drug was systematically investigated by Miller et al. comparing never-dried BNC with freeze-dried one [46]. Freeze-dried samples showed a lower uptake capacity for albumin than native BNC, which was found to be related to changes of the fibre network during the freeze- drying process. BNC was functionalised with the antiseptic drug octenidine, and its mechanical characteristics, biocom-patibility, and antimicrobial efficacy were investigated and concluded that it represents a ready-to-use wound dressing for the treatment of infected wounds, without losing its antibacterial activity over six months [45].

10.6 Synthesis of Giant Vesicles from Nanocellulose

Our group has showed for the first time that modification of nano-fibrillated cellulose (NFC) synthesised from the husk fibres of areca husk fibres (AHF) by controlled regio-selective amidation with oleylamine (OA) showed giant vesicular assembly [63]. The modified system (MNFC) with more than 66% OA content showed self-assembly into unilamellar vesicles of 2–5 μm diameters having wall thickness of 300–600 nm in tetrahydrofuran solution at 2.5 mg mL-1 (Figure 10.7). The formation of large vesicles in modified nanocellulose was attributed to the folding of MNFC into bilayers driven by

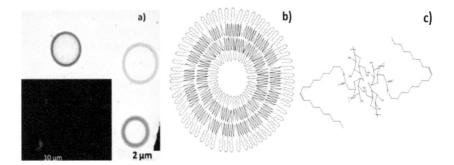

Figure 10.7 (a) Transmission electron microscope image with fluorescence microscopy image inset of self-assembled vesicles from modified nanofibrillated cellulose at 2.5 mg mL-1 in THF, (b) Cross sectional representation of the vesicle (blue lines indicate nanofibrils) and (c) typical hydrogen bonding interactions expected in the modified fibrils in the vesicle wall.

long cis-unsaturated aliphatic chains in polar aprotic solvents stabilised by hydrogen bonded interactions within the fibrils. These giant vesicles formed can have possible applications in therapeutics for topical applications.

10.7 Conclusions

Polysaccharide based self-assembled nanostructures such as vesicles are having immense potential as a green, nontoxic drug delivery system for therapeutic applications. The size and wall thickness of the vesicular assembly of long alkenyl chain modified nanocellulose can be controlled by controlling the regio-specific modification using oleyl amine. These structures are of the same thickness of the cell walls to mimic their transport properties. Nanocellulose, based polysaccharide self-assembly is expected to lead into various morphologies opening up a novel strategy for controlled and targeted drug delivery especially in cancer chemotherapy which are stimuli responsive with minimum side effects. It opened up new vistas for synthesis of green nano morphologies for biomedical applications especially in therapeutics.

References

[1] Abeer, M. M., Amin, M.C.I.M., Marti, M.; A review of bacterial cellulose-based drug delivery systems: their biochemistry, current approaches and future prospects, Journal of Pharmacy and Pharmacology; 2014, 66(8), 1047-1061.

[2] Agnihotri, S. A., Jawalkar, S.S., Aminabhavi, T. M.; Controlled release of cephalexin through gellan gum beads: effect of formulation parameters on entrapment efficiency, size, and drug release, Eur. J. Pharm. Biopharm.; 2006, 63, 249–261.

[3] Agnihotri, S. A., Aminabhavi, T. M.; Development of novel interpenetrating network gellan gum-poly(vinyl alcohol) hydrogel microspheres for the controlled release of carvedilol, Drug Dev. Ind. Pharm.; 2005, 31, 491–503.

[4] Agnihotri, S. A., Jawalkar, S. S., Aminabhavi, T. M.; Controlled release of cephalexin through gellan gum beads: effect of formulation parameters on entrapment efficiency, size, and drug release, Eur. J. Pharm. Biopharm.; 2006, 63, 249–261.

[5] Aliabadi, M. H., Lavasanifar, A; Polymeric micelles for drug delivery, Expert Opin. Drug Deliv.; 2006, 3(1), 139-162.

[6] Ariga, K.; Lvov, Y. M.; Kawakami, K.; Ji, Q.; Hill, J. P.; Layer-by-layer self-assembled shells for drug delivery., Advanced Drug Delivery Reviews; 2011, 63, 762–771.

[7] Aumelas, A., Serrero, A., Durand, A., Dellacherie, E., Leonard, M.; Nanoparticles of hydrophobically modified dextrans as potential drug carrier systems. Colloids Surfaces B, 2007,59, 74–80.

[8] Badwaik, H.R., Giri, T.K., Nakhate, K.T., Kashyap, P., Tripathi, D.K.; Xanthan gum and its derivatives as a potential bio-polymeric carrier for drug delivery system. Curr. Drug Deliv., 2013, 10, 587–600.

[9] Bera, H., Boddupalli, S., Nandikonda, S., Kumar, S., Nayak, A.K.; Alginate gel-coated oil- entrapped alginate-tamarind gum-magnesium stearate buoyant beads of risperidone. Int. J. Biol. Macromol.; 2015,78, 102–111.

[10] Bhatt, A. D., Ranade, A. A., A Retrospective Observational Study of Efficacy and Safety of Genexol-PM, A Novel Cremophor-Free, Polymeric Micelle Formulation of Paclitaxel, in Patients with Solid Tumours, Ann. Oncol.; 2016, 27.

[11] Brosnan, S. M.; Schlaad, H.; Antonietti, M.; Aqueous Self-Assembly of Purely Hydrophilic Block Copolymers into Giant Vesicles, Angew. Chem. Int. Ed.; 2015, 54, 9715 –9718.

[12] Cafaggi, S., Russo, E., Stefani, R., Parodi, B., Caviglioli, G., Sillo, G., Bisio, A., Aiello, C., Viale, M.; Preparation, characterisation and preliminary antitumour activity evaluation of a novel nanoparticulate system based on a cisplatin-hyaluronate complex and N-trimethyl chitosan, Invest. New Drugs; 2011, 29, 443–455.

[13] Choisnard L, Geze A, Putaux J, Wong Y, Wouessidjewe D. Nanoparticles of β-cyclodextrin esters obtained by self-assembling of biotransesterified β-cyclodextrins. Biomacromolecules; 2006, 7, 515–520.

[14] Coviello, T., Dentini, M., Rambone, G., Desideri, P., Carafa, M., Murtas, E., Riccieri, F.M., Alhaique, F.; A novel co-crosslinked polysaccharide: studies for a controlled delivery matrix. J. Control. Release.; 1999, 60, 287–295.

[15] Coviello, T., Matricardi, P., Marianecci, C., Alhaique, F., Polysaccharide hydrogels for modified release formulations. J. Control. Release, 2007, 119, 5–24.

[16] Cui, W., Junbai Li, J., Decher, G.; Self-Assembled Smart Nanocarriers for Targeted Drug Delivery, Adv. Mater.; 2016, 28, 1302–1311.

[17] Deshpande, N. U., Jayakannan, M., Cisplatin-Stitched Polysaccharide Vesicles for Synergistic Cancer Therapy of Triple Antagonistic Drugs, *Biomacromolecules*, 2017, 18(1),113–126.

[18] Douer, D; Efficacy and Safety of Vincristine Sulfate Liposome Injection in the Treatment of Adult Acute Lymphocytic Leukemia, Oncologist; 2016, 21, 840-847.

[19] Fialho, A.M., Moreira, L.M., Granja, A.T., Popescu, A.O., Hoffmann, K., Sá-Correia, I.; Occurrence, production, and applications of gellan: current state and perspectives. Appl. Microbiol. Biotechnol; 2008, 79, 889–900.

[20] García-González, C.A.; Alnaief, M.; Smirnova, I.; Polysaccharide-based aerogels—Promising biodegradable carriers for drug delivery systems, Carbohydrate Polymers; 2011, 86, 1425– 1438.

[21] Gaucher, G.; Satturwar, P., Jones, M. C., Furtos, A., Leroux, J. C.; Polymeric micelles for oral drug delivery, European Journal of Pharmaceutics and Biopharmaceutics; 2010, 76, 147–158.

[22] Gavini, E., Hegge, A. B., Rassu, G., Sanna, V., Testa, C., Pirisino, G., Karlsen, J., Giunchedi, P.; Nasal administration of carbamazepine using chitosan microspheres: in vitro/in vivo studies, Int. J. Pharm.; 2006, 307, 9–15.

[23] Gavini, E., Rassu, G., Sanna, V., Cossu, M., Giunchedi, P.; Mucoadhesive microspheres for nasal administration of an antiemetic drug, metoclopramide: *in-vitro/ex-vivo* studies. J. Pharm. Pharmacol.;2005,57, 287–294.

[24] George, M., Abraham, T.E.; pH sensitive alginate-guar gum hydrogel for the controlled delivery of protein drugs. Int. J. Pharm.; 2007, 335, 123–129.

[25] Heinze, T., Liebert, T., Heublein, B., Hornig, S., Functional polymers based on dextran. Adv. Polym. Sci. 2006,205–206, 199–291.

[26] Houga, C., Giermanska, J., Lecommandoux, S., Borsali, R., Taton, D., Gnanou, Y., Meins J. L.; Micelles and Polymersomes Obtained by Self-Assembly of Dextran and Polystyrene Based Block Copolymers, Biomacromolecules; 2009, 10, 32–40.

[27] Huang, L., Chen, X., Nguyen, T. X., Tang, H., Zhang, L., Yang,G.; Nano-cellulose 3D- networks as controlled-release drug carriers, Mater. Chem. B; 2013, 1,2976.

[28] Jain, A., Jain, S. K., Ganesh, N., Barve, J., Beg, A. M.; Design and development of ligand-appended polysaccharidic nanoparticles for the delivery of oxaliplatin in colorectal cancer, Nanomedicine; 2009, 6, 179–190.

[29] Kassem, A.A., Farid, R.M., Issa, D.A., Khalil, D.S., Abd-El-Razzak, M.Y., Saudi, H.I., Eltokhey, H.M., El-Zamarany, E.A.; Development of mucoadhesive microbeads using thiolated sodium alginate for intrapocket delivery of resveratrol. Int. J. Pharm.; 2015, 487, 305–313.

[30] Kim, B., B., Park, S. W., Hammond, P. T.; Hydrogen-Bonding Layer-by-Layer-Assembled Biodegradable Polymeric Micelles as Drug Delivery Vehicles from Surfaces, ACS nano; 2008, 2(2), 386–392.

[31] Kolakovic, R., Peltonen, L., Laukkanen A., Hirvonen, J., Laaksonen,T.; Nanofibrillar cellulose films for controlled drug delivery, European Journal of Pharmaceutics and Biopharmaceutics; 2012, 82, 308–315

[32] Kumar, C. G., Poornachandra, Y., Pombala, S.; Therapeutic Nanomaterials: From A Drug Delivery Perspective, In Nanostructures for Drug Delivery, Micro and nano Technologies; Chapter 1, Elsevier, 2017, pp1-6.

[33] Li, H., Deng, Y., Liu, B., Ren, Y., Liang, J., Qian, Y., Qiu, X., Li, C., Zheng, D.; Preparation of nanocapsules via self-assembly of kraft lignin: A totally green process with renewable resources, ACS Sustainable Chemistry & Engineering; 2016, 1-29.

[34] Li, M., Luo, Z., Zhao, Y.; Self-Assembled Hybrid Nanostructures: Versatile Multifunctional Nanoplatforms for Cancer Diagnosis and Therapy, Chem. Mater.; 2018, 30, 25-53.

[35] Li, Y. L., Zhu, L., Liu, Z., Cheng, R., Meng, F., Cui, J.H., Ji, S.J., Zhong, Z., Reversibly stabilized multifunctional dextran nanoparticles efficiently deliver doxorubicin into the nuclei of cancer cells. Angew. Chem. Int. Ed. Engl. 2009, 48, 9914–9918.

[36] Lin H. L., Chen, X., Nguyen, T.X., Tang, H., Zhang, L., Yang,G.; Nanocellulose 3D- networks as controlled-release drug carriers, J. Mater. Chem. B, 2013, 1, 2976–2984.

[37] Lin, N., Dufresne. A.; Nanocellulose in biomedicine: Current status and future prospect, European Polymer Journal; 2014, 59, 302-325.

[38] Lin, Y. H., Mi, F. L., Chen, C. T., Chang, W. C., Peng, S. F., Liang, H. F., Sung, H. W.; Preparation and characterization of nanoparticles shelled with chitosan for oral insulin delivery, Biomacromolecules; 2007, 8, 146–152.

[39] Liu, Y., Sui, Y., Liu,C., Liu,C.,Wu,M., Li,B., Li, Y.; A physically crosslinked polydopamine/nanocellulose hydrogel as potential versatile vehicles for drug delivery and wound healing. Carbohydrate Polymers.;2018, 188,27-36.

[40] Liu, Y., Sun, J., Zhang, P., He, Z.; Amphiphilic Polysaccharide-Hydrophobicized Graft Polymeric Micelles for Drug Delivery Nanosystems, Current Medicinal Chemistry, 2011, 18, 2638-2648.

[41] Liu, Z., Jiao, Y., Wang, Y., Zhou, C., Zhang, Z.,; Polysaccharides-based nanoparticles as drug delivery systems, Advanced Drug Delivery; 2008, 60, 1650–1662.

[42] Lorenzo, C. A., Fernandez, B, B., Puga, A. M., Concheiro, A.; Crosslinked ionic polysaccharides for stimuli-sensitive drug delivery, Advanced Drug Delivery Reviews; 2013, 65(9), 1148-71.

[43] Matricardi, P., Meo, C.D., Coviello, T., Alhaique, F., Recent advances and perspectives on coated alginate microspheres for modified drug delivery. Expert Opin. Drug Deliv; 2008, 5, 417–425.

[44] Mehvar, R., Dextrans for targeted and sustained delivery of therapeutic and imaging agents. J. Control. Release, 2000, 69, 1–25.

[45] Mooranian, A., Negrulj, R., Mikov, M., Golocorbin-Kon, S., Arfuso, F., Al-Salami, H.; Novel chenodeoxycholic acid-sodium alginate matrix in the microencapsulation of the potential antidiabetic drug, probucol: an in vitro study. J. Microencapsul.; 2015, 32, 589–597.

[46] Moritz, S., Wiegand, C., Wesarg, F., Hessler, N., Müller, F. A., Kralisch, D., Hipler, U. C., Fischer, F.; Active wound dressings based on bacterial nanocellulose as drug delivery system for octenidine, International Journal of Pharmaceutics; 2014, 471, 45–55.

[47] Muller, A., Ni,Z., Hessler,N., Wesarg, F., Mu Ller, F. A., Kralisch,D., Fischer A.; The Biopolymer Bacterial Nanocellulose as Drug Delivery System: Investigation of Drug Loading and Release using the

Model Protein Albumin, Journal of Pharmaceutical Sciences; 2013, 102, 579–592.

[48] Mura, S., Nicolas,J., Couvreur, P., Stimuli-Responsive Nanocarriers For Drug Delivery, Nature Materials; 2013, 12, 991-1003.

[49] Myrick, J. M., Vendraa V. K., Krishnan, S.; Self-assembled polysaccharide nanostructures for controlled-release applications, Nanotechnol; 2014, 3(4): 319–346.

[50] Naa, K., Leea, T. B., Parka, K. H., Shina, E. K., Leeb, Y.B., Choi, H. K.; Self-assembled nanoparticles of hydrophobically-modified polysaccharide bearing vitamin H as a targeted anti-cancer drug delivery system, Eur. J. Pharm. Sci.; 2003, 18, 165–173.

[51] Nayak, A. K., Pal, D.; Development of pH-sensitive tamarind seed polysaccharide-alginate composite beads for controlled diclofenac sodium delivery using response surface methodology. Int. J. Biol. Macromol.; 2011., 49, 784–793.

[52] Osma'ek, T., Froelich, A., Tasarek, S.; Application of gellan gum in pharmacy and medicine, Int. J. Pharm.; 2014, 466, 328–340.

[53] Paliwal, R., Paliwal, S. R., Agrawal, G. P., Vyas, S. P.; Chitosan nanoconstructs for improved oral delivery of low molecular weight heparin: in vitro and in vivo evaluation, Int. J. Pharm.; 2012, 422, 179–184.

[54] Pastori, G., Guolo, F., Guardo, D., Minetto, P., Clavio, M., Miglino, M., Giannoni, L., Coviello, E., Ballerini, F., Galaverna, F., Kunki, A., Colombo, N.,Grasso, R., Lemoli, R. M., Gobbi, M.; Fludarabine, Cytarabine, Daunoxome Plus Dasatinib has High Efficacy with an Acceptable Toxicity Profile as Either Consolidation or Salvage Regimen in Adult Philadelphia Positive Acute Lymphoblastic Leukemia Patients. Blood; 2015, 126, 4908.

[55] Pillay, V., Dangor, C. M., Govender, T., Moopanar, K. R., Hurbans, N.; J. Microencapsulation Micro and Nano Carriers; 1998, 15(2), 215-216.

[56] Plackett, D. V., Letchford, K., Jackson, J. K., Burt, H. M; A review of nanocellulose as a novel vehicle for drug delivery, Nordic Pulp & Paper Research Journal; 2014, 29 (1), 105-118.

[57] Pramod, P. S., Deshpande, N. U., Jayakanna, M.; Real-Time Drug Release Analysis of Enzyme and pH Responsive Polysaccharide Nanovesicles. J. Phys. Chem. B; 2015, 119, 10511-10523.

[58] Qi, W., Wang, A. H., Yang, Y., Du, M. C., Bouchu, M. N., Boullangerd, P., Li, J.; The lectin binding and targetable cellular uptake of lipid-coated polysaccharide microcapsules, J. Mater. Chem. 2010, 20, 2121-2127.

[59] Rehm, B.H.A., Alginates: Biology and Applications. Springer-Verlag, Berlin, Heidelberg, 2009.

[60] Schatz, C., Louguet, S. P., Meins, J. O. L., Lecommandoux, S. B.; Polysaccharide-block- polypeptide Copolymer Vesicles: Towards Synthetic Viral Capsids Angew. Chem. Int. Ed.; 2009, 48, 2572 –2575.

[61] Shutava, T.G., Balkundi, S.S., Vangala, P., Steffan, J.J., Bigelow, R.L., Cardelli, J.A., O'Neal, D.P., Lvov, Y. M.; Layer-by-layer-coated gelatin nanoparticles as a vehicle for delivery of natural polyphenols, ACS Nano; 2009, 3, 1877–1885.

[62] Sinha, V.R., Kumria, R., Polysaccharides in colon-specific drug delivery, International Journal of Pharmaceutics; 2001, 244, 19–38.

[63] Sofias, A. M., Dunne, M., Storm, G., Allen, C.; The Battle of ?Nano? Paclitaxel, Adv. Drug Delivery Rev; 2017,122, 20-30.

[64] Soman, S., Chacko, A.S., Prasad, V.S., Anju, P., Surya, B. S., Vandana, K.; Self-assembly of oleylamine modified nano-fibrillated cellulose from areca husk fibers into giant vesicles, Carbohydrate Polymers; 2018,182, 69–74.

[65] Song, J., Zhou, J., Duan, H.; Self-Assembled Plasmonic Vesicles of SERS-Encoded Amphiphilic Gold Nanoparticles for Cancer Cell Targeting and Traceable Intracellular Drug Delivery, J. Am. Chem. Soc.; 2012, 134, 13458-13469.

[66] Song, Y., Zhang, L., Gan, W., Zhou, J., Zhang, L.; Self-assembled micelles based on hydrophobically modified quaternized cellulose for drug delivery, Colloids and Surfaces B: Biointerfaces; 2011, 83, 313–320.

[67] Sonoda, A., Nitta, N., Nitta-Seko, A., Ohta, S., Takamatsu, S., Ikehata, Y., Nagano, I., Jo, J., Tabata, Y., Takahashi, M., Matsui, O., Murata, K.; Complex comprised of dextran magnetite and conjugated cisplatin exhibiting selective hyperthermic and controlled-release potential, Int. J. Nanomed.; 2010, 5, 499–504.

[68] Sridhar, U., Pramod, P. S., Jayakannan, M., Creation of dextrin vesicles and their loading- delivering capabilities, RSC Adv., 2013, 3, 21237.

[69] Urban, C; Treatment of Metastatic Breast Cancer: New Data from SABCS 2016 Emphasize the Importance of Eribulin (Halaven), Breast Care; 2017, 12, 64-65.

[70] Varshosaz, J., Emami, J., Tavakoli, N., Fassihi, A., Minaiyan, M., Ahmadi, F., Dorkoosh, F., Synthesis and evaluation of dextran-budesonide conjugates as colon specific prodrugs for treatment of ulcerative colitis. Int. J. Pharm. 2009, 365, 69–76.

[71] Vyas, S., Trivedi, P., Chaturvedi, S.C., Dextran-etodolac conjugates: synthesis, in vitro and in vivo evaluation. Acta Pol. Pharm. 2009, 66, 201–206.

[72] Wittayaareekul, S., Kruenate, J., Prahsarn, C.; Preparation and in vitro evaluation of mucoadhesive properties of alginate/chitosan microparticles containing prednisolone. Int. J. Pharm.; 2006, 312, 113–118.

[73] Xing, L., Dawei, C., Liping, X., Rongqing, Z.; Oral colon-specific drug delivery for bee venom peptide: development of a coated calcium alginate gel beads-entrapped liposome, J. Control. Release; 2003, 93, 293–300.

[74] Yang, Y., Wang, S., Wang, Y., Wang, X., Wang, C., Chen, M.; Advances in self-assembled chitosan nanomaterials for drug delivery, Biotechnology Advances; 2014, 32(7), 1301-1316.

[75] Zhang, A., Zhang, Z., Shi, F., Xiao, C., Ding, J., Zhuang, X., He, C., Chen, L., Chen, X.; Redox-Sensitive Shell-Crosslinked Polypeptideblock-Polysaccharide Micelles for Efficient Intracellular Anticancer Drug Delivery, Macromol. Biosci.; 2013,13(9), 1249-1258.

[76] Zhang, L., Gu, F. X., Chan, J.M., Wang, A. Z., Langer, R. S., Farokhzad, O. C; Nanoparticles in Medicine: Therapeutic Applications and Developments, Clinical Pharmacology & Therapeutics; 2008, 83(5), 761-769.

[77] Zu, Y., Wang, D., Zhao, X., Jiang, R., Zhang, Q., Zhao, D., Li, Y., Zu, B., Sun, Z.; A novel preparation method for camptothecin (CPT) loaded folic acid conjugated dextran tumor-targeted nanoparticles, Int. J. Mol. Sci.; 2011, 12, 4237–4249.

11

Antimicrobial Effects of Biosynthesised Silver Nanoparticles Using *Pimenta Dioica* Leaf Extract

Reshma R. Pillai, P. B. Sreelekshmi and A. P. Meera*

Research and Post Graduate Department of Chemistry & Polymer Chemistry, KSMDB College, Sasthamcotta, Kollam, Kerala, India
E-mail: apmeera@gmail.com; meeradbc@gmail.com
*Corresponding author

In this study, we report an extracellular rapid biosynthesis of silver nanoparticles (AgNPs) from $AgNO_3$ using aqueous leaf extract of *Pimenta dioica*. The main purpose of this study is to minimise the hazards of chemical synthesis and to develop an environment friendly method for the synthesis of silver nanoparticles. Here the aqueous leaf extract of *Pimenta dioica* was used as a reducing medium for Ag^+ ions of silver nitrate to metallic Ag. The synthesised nanoparticles were characterised by FT-IR, UV-Visible, SEM and TEM. The spherically shaped AgNPs are found to be in the size range of 20-35 nm. The antibacterial activity of the silver nanoparticles was evaluated using four types of bacteria, *Staphylococcus aureus*, *Escherichia coli, Klebsiella pneumonia* and *Pseudomonas aeroginosa*. The antifungal activity was studied with the zone of inhibition produced by two pathogenic fungi *Aspergillus niger and Candida albicans*. The biosynthesised silver nanoparticles show effective antimicrobial activity.

11.1 Introduction

Nanotechnology is an important field of modern research dealing with synthesis, strategy and manipulation of particle's structure ranging from approximately 1 to 100 nm in size [1]. Within this size range all the properties

217

(chemical, physical and biological) changes in fundamental ways of both individual atoms/molecules and their corresponding bulk. Novel applications of nanoparticles and nanomaterial are growing rapidly on various fronts due to their completely new or enhanced properties based on size, their distribution and morphology. It is swiftly gaining renovation in a large number of fields such as health care, cosmetics, biomedical, food and feed, drug-gene delivery, environment, health, mechanics, optics, chemical industries, electronics, space industries, energy science, catalysis, light emitters, single electron transistors, nonlinear optical devices and photo-electrochemical applications.

The nanoparticles used for all the aforesaid purposes, the metallic nanoparticles considered as the most promising as they contain remarkable antibacterial properties due to their large surface area to volume ratio, which is of interest for researchers due to the growing microbial resistance against metal ions, antibiotics and the development of resistant strains [2]. Among the all noble metal nanoparticles, silver nanoparticles (AgNPs) are an arch product from the field of nanotechnology which has gained boundless interests because of their unique properties such as chemical stability, good conductivity, catalytic and most important antibacterial, anti-viral, antifungal in addition to anti-inflammatory activities which can be incorporated into composite fibres, cryogenic superconducting materials, cosmetic products, food industry and electronic components [3, 4, 5]. For biomedical applications; being added to wound dressings, topical creams, antiseptic sprays and fabrics, silver functions' as an antiseptic and displays a broad biocidal effect against microorganisms through the disruption of their unicellular membrane thus disturbing their enzymatic activities. These silver nanoparticles are being successfully used in the cancer diagnosis and treatment as well [6–7].

Generally, nanoparticles are prepared by a variety of chemical and physical methods which are quite expensive and potentially hazardous to the environment which involve use of toxic and perilous chemicals that are responsible for various biological risks [8]. The development of biologically-inspired experimental processes for the synthesis of nanoparticles is evolving into an important branch of nanotechnology. Nowadays green synthesis of silver nanoparticles is an emerging area as it is nontoxic, cost effective and environment friendly. Green plants contain several biomolecules such as alkaloids, flavonoids and terpenes, which act as reducing as well as capping agents [9]. Although many studies have been reported on the synthesis of AgNPs using various plant extracts like *Calotropis procera [10], Achyranthes aspera [11], Bauhinia acuminata [12],* to the best our knowledge not much work has

Figure 11.1 Picture of (a) *Sarvasugandhi* leaves and (b) powdered form.

been reported using *Pimenta dioica* leaf extract [13]. *Pimenta dioica* which is commonly known as 'Sarvasugandhi' belongs to the family 'Myrteace'. It has anti-microbial, medicinal, insecticidal, anti-oxidant and deodorant properties. It is rich in potassium, manganese, iron, copper, selenium and magnesium. It is used for indigestion (dyspepsia), intestinal gas, abdominal pain, heavy menstrual periods, vomiting, diarrhoea, fever, colds, high blood pressure, diabetes, obesity etc. The present study aims to investigate the antimicrobial activities of silver nanoparticles synthesised via green route using *Pimenta dioica* leaf extract.

11.2 Experimental

The fresh leaves of 'Sarvasugandhi' *(Pimenta dioica)* as shown in Figure 11.1a were collected from the plant species. The leaves were rinsed thoroughly first with tap water followed by distilled water to remove all the dust and unwanted visible particles and dried at room temperature then sun-dried to remove the residual moisture. The dried leaves were ground to fine powder (Figure 11.1b).

11.2.1 Materials Used

11.2.1.1 Preparation of the leaf extract

30 gram of the dried *Pimenta dioica* leaves were powdered and boiled in a 250 ml glass beaker along with 200 ml of de-ionised water for 15 minutes. After boiling, the colour of the aqueous solution is changed from watery to yellowish colour. The aqueous extract was separated by filtration with Whatman No.1 filter paper to remove particulate matter and to get clear solutions which were then refrigerated (4 °C) the leaf extract stored to be used for the bio synthesis of silver nanoparticles from silver nitrate solution.

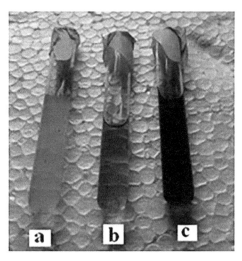

Figure 11.2   Figure showing (a) AgNO₃ solution, (b) leaf extract and (c) silver nanoparticles.

11.2.1.2 Synthesis of silver nanoparticles

About 40 ml of *Pimenta* leaf extract was added to 100 ml 0.002 M aqueous AgNO₃ solution with stirring magnetically at room temperature. The yellow colour of the mixture of silver nitrate and *Pimenta* leaf extract at zero min to reaction time changed very fast at room temperature after 15 min to a dark brown suspended mixture (Figure 11.2). This indicates that the *Pimenta* leaf extract speeds up the biosynthesis of AgNPs. The silver nanoparticles obtained by *Pimenta* leaf extract was centrifuged for 5 min.

11.2.2 Characterisation Techniques of Nanoparticles

11.2.2.1 Fourier transforms infrared spectroscopy (FTIR)

The synthesised silver nanoparticles were characterised by FTIR. The spectra were recorded on a 'Thermo Nicolet, Avatar 370' FTIR spectrometer in the wave number range 400–4000cm^{-1} by KBr disc method. The spectral analysis was done in Sophisticated Analytical Instrument Facility, STIC, CUSAT Cochin.

11.2.2.2 UV-visible spectroscopy

In UV-visible region of the electromagnetic spectrum, molecules undergo electronic transitions. The reduction of the Ag+ ions by the supernatant of the test plant extracts in the solutions and formation of silver nanoparticles were characterised by UV-visible spectroscopy. Electronic spectra are recorded on 'Perkin Elmer' UV–visible spectrometer in the wavelength range 400–450 nm.

11.2.2.3 Morphological analysis

Scanning electron microscopy (SEM) and Transmission electron microscopy (TEM) are the common methods used for the surface and morphological characterisation at the nanometre to micrometer scale. SEM can provide morphological information on the submicron scale and elemental information at the micron scale. TEM has a much higher resolution compared with the SEM. So, to know the surface morphology of the prepared nanoparticle sample same is used and TEM is useful to know the exact size and shape of nanoparticles.

11.2.3 Biological Activity of the Nanoparticles

11.2.3.1 Antibacterial activity

Antibacterial properties were performed at CEPC, Kollam and Bio-genix, Thiruvananthapuram. Antibacterial activities of the synthesised silver nanoparticles have been carried out against the pathogenic bacteria, *Escherichia coli* and *Staphylococcus Aureus* using Muller Hinton Agar medium by well diffusion method.

Filter paper disc diffusion technique was applied for determining the antibacterial activity against *E. coli* and *S. aureus*. A sterile non-toxic swab was dipped on a wooden applicator into the standardised inoculum and the soaked swab was rotated firmly against the upper side wall of the tube to express the excess fluid; streaked the entire agar surface of the plate with the swab three times, turning the plates at 60 angles between each streaking; allowed the inoculum to dry for 5–15 minutes with lid in place; applied the disc (Hi media sterile 6mm disc) impregnated with the sample, approximately 30 μl, using aseptic technique; then placed the discs with centres at least 24 mm apart; incubated immediately at 37 °C; and examined after 16–18 hours or later if necessary. The zone showing complete inhibition is measured and the diameters of the zones to the nearest millimetre are recorded. Discs soaked in pure solvent; Di methyl sulphoxide (DMS) were used as control.

Petriplates containing 20 ml Muller Hinton Agar Medium were seeded with bacterial culture of *Pseudomonas aeroginosa* and *Klebsiella pneumoniae* (growth of culture adjusted according to McFards Standard, 0.5%). Wells of approximately 10mm was bored using a well cutter and different concentrations of sample such as 25 μL, 50 μL, 100 μL were added. The plates were then incubated at 37°C for 24 hours. The antibacterial activity was assayed by measuring the diameter of the inhibition zone formed around the well (NCCLS, 1993). Streptomycin was used as a positive control.

11.2.3.2 Antifungal activity

In order to access the biological significance and ability of the sample, the antifungal activity was determined by Agar well diffusion method. Potato Dextrose agar plates were prepared and overnight grown species of fungus, *Aspergillus niger* (ATCC 16404) and *Candida albicans* (ATCC 10231) was swabbed. Wells of approximately 10 mm was bored using a well cutter and samples of different concentration was added; the zone of inhibition was measured after overnight incubation and compared with that of standard antimycotic (Clotrimazole).

11.3 Results and Discussion

11.3.1 Characterisation of Silver Nanoparticles

11.3.1.1 UV-visible spectroscopy

It is recognised that UV–VIS spectroscopy could be used to examine size and shape controlled nanoparticles in aqueous suspensions [14]. It is well known that silver nanoparticles exhibit yellowish brown colour in aqueous solution due to excitation of surface Plasmon vibrations in silver nanoparticles [15]. As the extract was mixed in the aqueous solution of the silver ion complex, it started to change the colour from colourless to yellowish brown due to reduction of silver ion which indicated formation of silver nanoparticles.

Absorption of light at wavelength of 400–450 nm is due to collective oscillation of surface electrons in AgNPs [16] (Figure 11.3). The broadening of peak indicated that the particles are polydispersed.

11.3.1.2 FT-IR spectral studies

The IR spectrum of Ag nanoparticles (Figure 11.4) shown band at 3425 cm^{-1} which represents the O-H stretching of hydrogen bonded alcohols and phenols [17, 18]. The peak located at 1630 cm^{-1} could be assigned to C=O stretching or amide bending [19, 20]. The peak at 1384 cm^{-1} assigned to nitro N-O bending [21] and a peak at 1092 cm^{-1} to C-O-C stretching aromatic ring [22].

FTIR spectrum of Ag nanoparticles suggests that Ag nanoparticles are surrounded by different organic molecules such as terpenoids, alcohols,

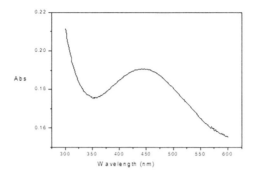

Figure 11.3 UV-Vis spectra of silver nanoparticles of *Pimenta dioica.*

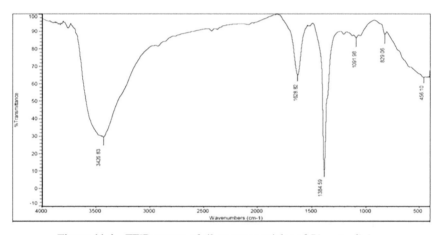

Figure 11.4 FTIR spectra of silver nanoparticles of *Pimenta dioica.*

ketones, aldehydes, carboxylic acids etc. The FT-IR spectrum clearly indicates the presence of the residual plant extract in the sample as a capping agent to the silver nanoparticles.

11.3.2 Morphological Analysis

11.3.2.1 Scanning electron microscopy (SEM)

The synthesised silver nanoparticles were characterised by SEM. The SEM micrographs are shown in Figure 11.5. The dispersion of nanoparticles can be clearly seen in the micrographs. The presence of the residual plant extract in the sample as a capping agent to the nanoparticles can also be seen in the images.

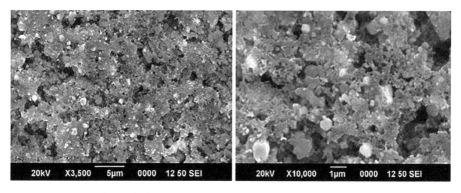

Figure 11.5   SEM micrographs of the Ag nanoparticles.

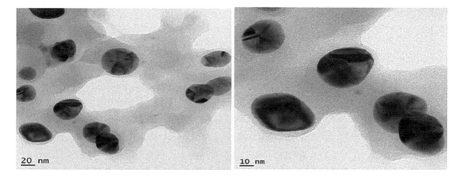

Figure 11.6   TEM images of the Ag nanoparticles.

11.3.2.2 Transmission electron microscopy (TEM)

The nano morphology of the nanoparticles was analyzed by TEM studies. The TEM images of the silver nanoparticles are shown in Figure 11.6.

Transmission Electron Microscopy (TEM) results showed particles with spherical shape surrounded by biological molecules, which prevent Ag nanoparticles from aggregation. The average size of the nanoparticles is found to be in the 20–35 nm range.

11.3.3 Biological Activity of the Complexes

The newly synthesised silver nanoparticles were screened in vitro for their antibacterial activity against bacteria: *Staphylococcus aureus* and *E. coli* (Figure 11.7). Dimethyl sulphoxide (DMS) was used as control. Figure 11.8a

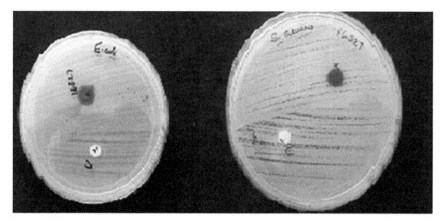

Figure 11.7 Antibacterial activity of the silver nanoparticles against *E. coli* and *Staphylococcus aureus.*

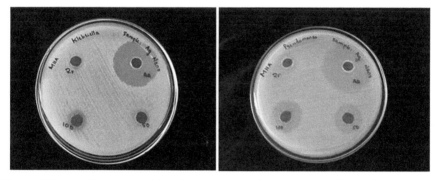

Figure 11.8 Antibacterial activity of silver nanoparticles against (a) *Klebsiella pneumonia* and (b) *Pseudomonas aeroginosa.*

and b showed the antibacterial activity against *Klebsiella pneumoniae* and *Pseudomonas aeroginosa*. The zone inhibition of bacterial growth was measured in mm. The anti bacterial activity can be assessed based on the diameter of the inhibition zone. It was observed that the nanoparticles show good antimicrobial activity towards Staphylococcus aureus (15 mm) and E-coli bacteria (8 mm). The antibacterial activity of *K. pneumoniae* and *P. aeroginosa* are shown in Table 11.1. Figure 11.9a and b showed the antifungal activity of silver nanoparticles against *Candida albicans* and *Aspergillus niger* (Table 11.2). As the concentration of silver nanoparticles increases the antimicrobial activity also increases.

Table 11.1 Results showing antibacterial activity of silver nanoparticles against (a) *Klebsiella pneumonia* and (b) *Pseudomonas aeroginosa.*

Organism	Concentration (μl) of Nano Ag	Zone of Inhibition
Klebsiella pneumonia	25	10
	50	11
	100	14
	Streptomycin (10μl)	33
Pseudomonas aeroginosa	25	Nil
	50	19
	100	25
	Streptomycin (10μl)	35

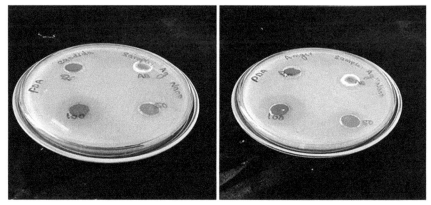

Figure 11.9 Antifungal activity of silver nanoparticles against (a) *Candida albicans* and (b) *Aspergillus niger.*

Table 11.2 Results showing antifungal activities of Ag nanoparticles.

Organism	Concentration (μl) of Nano Ag	Zone of Inhibition
Aspergillus niger	25	11
	50	14
	100	21
	Clotrimazole (10 μl) (Std.)	13
Candida albicans	25	Nil
	50	11
	100	20
	Clotrimazole (10μl) (Std.)	12

11.4 Conclusions

Silver nanoparticles were synthesised by green route using 'Sarvasugandhi' plant extract. The nanoparticles were characterised by IR and UV spectroscopy. FTIR and UV-Vis analysis confirmed the reduction of Ag (I) ions to Ag (0) which is supposed through the plant extract as capping agents i.e. the phytochemical constituents which are acting as the reducing agents. The morphology of the particles was assessed by SEM and TEM. The antibacterial activity of the silver nanoparticles was evaluated using four types of bacteria, *Staphylococcus aureus*, *Escherichia coli*, *Klebsiella pneumonia*, *Pseudomonas aeroginosa* and antifungal activity against *Candida albicans* and *Aspergillus niger*. The biosynthesised silver nanoparticles showed effective antibacterial and antifungal activities.

Acknowledgements

We thankfully acknowledge UGC, New Delhi, for the financial support (1741-MRP/14-15/KLKE012/UGC-SWRO). The authors are thankful to STIC Cochin, CEPC Kollam, Kerala, and Biogenix, Trivandrum, Kerala, for their support to carry out this work.

References

[1] Saware K, Sawle B, Salimath B, Jayanthi K, Abbaraju V, *Int. J. Res Eng. Tech.*,03 (2014).

[2] Khalil K.A, Fouad H, Elsarnagawy T, Almajhdi F.N. *Int J Electrochem-Sci* 8 (2013) 3483–93.

[3] Ahmad A , Mukherjee P , Senapati S, Mandal D, Khan M.I , Kumar R , Sastry M. *Colloids Surf B: Biointerfaces* 28 (2003) 313–8.

[4] Klaus-Joerger T, Joerger R, Olsson E, Granqvist C. *Trends Biotechnol* 19 (2001) 15–20.

[5] Rajamani R.K, Kuppusamy S, Bellan C.S, Ravi P.H, Sagadevan P, *MOJ Toxicol*, 4(3) (2018), 103-109.

[6] Popescu M, Velea A, Lorinczi A. *J Dig Nanomater Bios* 5(4) (2010) 1035–40.

[7] Baruwati B, Polshettiwar V, Varma R.S. *Green Chem* 11 (2009) 926–30.

[8] Arshad M, Khan A, Farooqi Z.H, Usman M, Waseem M.A, Shah S.A, Khan M, *Materials Science-Poland*, 36(1), (2018) 21–26.

 [9] Thomas B, Vithiya B.S.M, Prasad T.A.A, Mohamed S.B , Magdalane C.M, Kaviyarasu K, Maaza M, *Journal of Nanoscienceand Nanotechnology*, 18, (2018) 1–9.

[10] Nipane S.V, Mahajan P.G, Gokavi G.S, *Int. J. on Recent and Innovation Trends in Computing and Commun.* 04 (2016).

[11] Vijayaraj R, Naresh Kumar K, Mani P, Senthil J, Dinesh Kumar G, Jayaseelan T, *Int. J. of Pharmacy and Therapeutics* 7(1) (2016) 42–48.

[12] Divya Sebastian, Fleming A.T, *World J. of Pharmacy and Pharmaceutical Sciences*, 6(4), (2017) 1889–1900.

[13] Geetha A.R, George E, Srinivasan A, Shaik J, *The Scientific World Journal* (2013).

[14] Wiley B.J, Im S.H, Li Z.Y, Mclellan J, Siekkenen A, Xia Y. *J Phys Chem B*, 110(32) (2006) 15666–15675.

[15] Shiv Shankar S, Rai A, Ahmad A, Sastry M. *J Colloid Interf Sci*, 275 (2004) 496–502.

[16] Chaudhuri S.K, Chandela S, Malodia L, *Nano Biomed.Eng.* 8(1) (2016) 1–8.

[17] Quiawan M.J, Billacura M, Canalita D, *Sci.Int. (Lahore),* 29(2) (2017) 13–17.

[18] Nakhjiri F, Mirhosseini M, Mozaheb M.A, *Nanomed. J.,* 4(2) (2017) 98–106.

[19] Abbasa Q, Saleemb M, Phulla A.R, Rafiqa M, Hassana M, Ki-Hwan Leeb, Sung-Yum Seoa, *IJPR* 16(2) (2017) 760–767.

[20] Radhika P.R, Loganathan P, Sampath Kumar R, *Int. J. of Res. in Pharmacology & Pharmacotherapeutics,* 5(4) (2016) 135–141.

[21] Ahmed S, Ikram S, *J.Nanomed Nanotechnol* 6 (2015) 309.

[22] Anandalakshmi K, Venugobal J, *Med Chem*, (2017) 7:7.

Index

About the Editors

Didier Rouxel is full Professor at the Institut Jean Lamour, Université de Lorraine, France, where he was the former head of the « Micro and NanoSystems » team and currently leads the « Piezoelectric polymer » group. Rouxel is graduated from the Ecole Supérieure des Sciences et Techniques de l'Ingénieur de Nancy, Vandoeuvre, France. He completed his Ph.D in Material Sciences and Engineering from the University of Nancy I, France in 1993. He was expert for the French Agency for Food, Environmental and Occupational Health & Safety on "Nanomaterials and Health" and "Member of the Year 2014" of the French Society of Nanomedicine.

Prof. Rouxel has been involved in particular in the spectroscopic analysis of inorganics and is vastly experienced in the analysis of polymer nanocomposite systems by almost all spectroscopic techniques. His major areas of interest include piezoelectric polymers, elastic properties of polymeric materials studied by Brillouin spectroscopy, development of polymer nanocomposite materials, development of micro-devices based on electro-active polymers, piezoelectric nanocrystals, microsensor development for surgery, shape memory alloy-piezoelectric device for energy harvesting, etc.

Dr. Praveen K. M. is an Assistant Professor of Mechanical Engineering at Muthoot Institute of Technology & Scince (MITS) Ernakulam, Kerala, India. He pursued his PhD in Materials Engineering at the University of South Brittany (Université de Bretagne Sud) – Laboratory IRDL PTR1, Research Center "Christiaan Huygens," in Lorient, France, in the area of coir-based polypropylene micro composites and nanocomposites. He has published an international article in Applied Surface Science (Elsevier), 2 book chapters respectively by Elsevier and Springer Publishers and has also presented poster and conference papers at national and international conferences. Dr.Praveen edited an excellent book on Non-Thermal Plasma Technology for Polymeric Materials: Applications in Composites, Nanostructured Materials, and Biomedical Fields with Elsevier and 6 books with Apple Academic Press, Inc., USA He also has worked with the Jozef Stefan Institute, Ljubljana, Slovenia; Mahatma Gandhi University, India; and the Technical University in Liberec, Czech Republic. He is quite active in academic networking with various research groups around globe and organising technical events in his area of expertise. His current research interests include polymer composites, plasma modification of polymers, polymer composites for neutron shielding applications, and nanocellulose. He is a life member of Nanoscience and Nanotechnology Society, India.

Dr. Indu Raj completed her PhD in Dentistry (Modern Medicine, Mahatma Gandhi University, Kerala, India). She is Faculty member for "Rebase-Post graduate teaching programme" conducted by Indian Prosthodontic Society-Kerala chapter and also in various CDEs. She presented papers and posters in National and International conferences.

Dr. Sandhya Gopalakrishnan completed her BDS from Govt. Dental College, Trivandrum with university IInd rank and MDS (Prosthodontics) from MS Ramaiah Dental College, Bangalore. She did her PhD (Dentistry) from Mahatma Gandhi University, Kottayam. She has 13 years' post-PG teaching experience in Prosthodontics, and is a popular teacher at both UG and PG level. She is a recognized question setter for the National Board of Examination (Dental), Government of India. She has given several presentations, and has publications, in national and international journals, to her credit and has contributed chapters to 2 books. She is a reviewer for many journals. She has received several awards for her scientific presentations. Currently, she is working in Govt. Dental College, Kottayam as Associate Professor of Prosthodontics.

Dr. Nandakumar Kalarikkal is Joint Director at International and Inter University Centre for Nanoscience and Nanotechnology and Associate Professor at School of Pure and Applied Physics and Director of International and Inter University Centre for Nanoscience and Nanotechnology of Mahatma Gandhi University, Kottayam,

Kerala, India. His research activities involve applications of nanostructured materials, laser plasma, phase transitions, etc. He is the recipient of research fellowships and associateships from prestigious government organizations such as the Department of Science and Technology and Council of Scientific and Industrial Research of Government of India. He has active collaboration with national and international scientific institutions in India, South Africa, Slovenia,

Canada, France, Germany, Malaysia, Australia and US. He has more than 160 publications in peer reviewed journals. He has also co-edited 15 books of scientific interest and co-authored many book chapters.

Professor Sabu Thomas is the Director, School of Energy Materials, founder Director of the International and Interuniversity Centre for Nanoscience and Nanotechnology and full professor of Polymer Science and Engineering at the School of Chemical Sciences of Mahatma Gandhi University, Kottayam, Kerala, India. He is an outstanding leader with sustained international acclaims for his work in Polymer Science and Engineering, Polymer Nanocomposites, Elastomers, Polymer Blends, Interpenetrating Polymer Networks, Polymer Membranes, Green Composites and Nanocomposites, Nanomedicine and Green Nanotechnology. Dr. Thomas's ground breaking inventions in polymer nanocomposites, polymer blends, green nano technological and nano-biomedical sciences, have made transformative differences in the development of new materials for automotive, space, housing and biomedical fields. Professor Thomas has received a number of national and international awards which including, Fellowship of the Royal Society of Chemistry, London, MRSI medal, Nano Tech Medal, CRSI medal, Distinguished Faculty Award, and Sukumar Maithy Award for the best polymer researcher in the country. He is in the list of most productive researchers in India and holds a position of No.5. Thomas has been conferred Honoris Causa (D.Sc) by the University of South Brittany, Lorient, France in 2015 and in May 2016 he was awarded Loyalty Award in "International Materials Technology Conference Exhibition. Professor Thomas has published over 700 peer reviewed research papers, reviews and book chapters. He has co-edited nearly 58 books and is the inventor of 5 patents. He has supervised 79 Ph.D theses and his H index is 78 with nearly 27184 citations. Prof. Thomas has delivered over 300 Plenary/Inaugural and Invited lectures in national/international meetings over 30 countries. He has established a state of the art laboratory at Mahatma Gandhi University in the area of Polymer Science and Engineering and Nanoscience and Nanotechnology.